GEOGRAPHY, TECHNOLOGY AND INSTRUMENTS OF EXPLORATION

T0251060

Studies in Historical Geography

Series Editor: Professor Robert Mayhew, University of Bristol, UK

Historical geography has consistently been at the cutting edge of scholarship and research in human geography for the last fifty years. The first generation of its practitioners, led by Clifford Darby, Carl Sauer and Vidal de la Blache presented diligent archival studies of patterns of agriculture, industry and the region through time and space.

Drawing on this work, but transcending it in terms of theoretical scope and substantive concerns, historical geography has long since developed into a highly interdisciplinary field seeking to fuse the study of space and time. In doing so, it provides new perspectives and insights into fundamental issues across both the humanities and social sciences.

Having radically altered and expanded its conception of the theoretical underpinnings, data sources and styles of writing through which it can practice its craft over the past 20 years, historical geography is now a pluralistic, vibrant and interdisciplinary field of scholarship. In particular, two important trends can be discerned. Firstly, there has been a major "cultural turn" in historical geography which has led to a concern with representation as driving historical-geographical consciousness, leading scholars to a concern with text, interpretation and discourse rather than the more materialist concerns of their predecessors. Secondly, there has been a development of interdisciplinary scholarship, leading to fruitful dialogues with historians of science, art historians and literary scholars in particular which has revitalised the history of geographical thought as a realm of inquiry in historical geography.

Studies in Historical Geography aims to provide a forum for the publication of scholarly work which encapsulates and furthers these developments. Aiming to attract an interdisciplinary and international authorship and audience, Studies in Historical Geography will publish theoretical, historiographical and substantive contributions meshing time, space and society.

Geography, Technology and Instruments of Exploration

Edited by

FRASER MACDONALD AND CHARLES W.J. WITHERS
University of Edinburgh, UK

Routledge
Taylor & Francis Group

LONDON AND NEW YORK

First published 2015 by Ashgate Publishing

Published 2016 by Routledge
2 Park Square, Milton Park, Abingdon, Oxon OX14 4RN
711 Third Avenue, New York, NY 10017, USA

First issued in paperback 2018

Routledge is an imprint of the Taylor & Francis Group, an informa business

British Library Cataloguing in Publication Data
A catalogue record for this book is available from the British Library

The Library of Congress has cataloged the printed edition as follows:
MacDonald, Fraser, 1972-
  Geography, technology and instruments of exploration / by Fraser MacDonald and Charles W. J. Withers.
    pages cm. -- (Studies in historical geography)
  Includes bibliographical references and index.
  ISBN 978-1-4724-3425-8 (hardback)
1. Discoveries in geography--History. 2. Historical geography. 3. Geography--Instruments--History. 4. Navigation--Technological innovations--History. I. Withers, Charles W. J. II. Title.
  G80.M28 2015
  910.28--dc23
                                                                              2015011162

ISBN 13: 978-1-138-54725-4 (pbk)
ISBN 13: 978-1-4724-3425-8 (hbk)

Contents

List of Figures and Tables

Figures

Tables

Notes on Contributors

Peter Collier is Editor of *Survey Review* and Vice Chair of the Commission on the History of Cartography of the International Cartographic Association.

Richard Dunn is Senior Curator and Head of Science and Technology at Royal Museums Greenwich. Between 2010 and 2015, he was part of the research project, 'The Board of Longitude 1714–1828: Science, Innovation and Empire in the Georgian World', a collaboration between Royal Museums Greenwich and the Department of History and Philosophy of Science at the University of Cambridge, funded by the Arts and Humanities Research Council.

Isla Forsyth is Lecturer in Cultural and Historical Geography at the University of Nottingham. Her research interests concern military geographies, science and technology studies, the desert, and more-than-human geographies. Her research aims to produce narratives that focus on the hitherto overlooked lives of humans and non-humans who have shaped the geographies of technology, conflict, and the militarisation of culture.

Rebekah Higgitt is Lecturer in History of Science at the University of Kent. Her research has focused on the history of scientific institutions, including the Board of Longitude, the Royal Observatory, Greenwich, and the British Association for the Advancement of Science, and on the relationship between science, government and wider publics. She was, until recently, Co-Investigator on an AHRC-funded project with the National Maritime Museum and University of Cambridge on the history of the Board of Longitude.

John McAleer is Lecturer in History at the University of Southampton. His work focuses on the British engagement with the wider world in the eighteenth and nineteenth centuries, situating the history of empire in its global and maritime contexts. His current research is addressing 'The Keys to India: Britain's Maritime Empire and the Indian Ocean World'.

Fraser MacDonald is Lecturer in Historical Geography at the University of Edinburgh. He is currently working on a book-length history of rocketry in the mid twentieth century. He co-edited *Observant States: Geopolitics and Visual Culture* (I.B.Tauris, 2010) with R. Hughes and K. Dodds.

A.D. Morrison-Low is Principal Curator, Science, at National Museums Scotland. She has published widely on aspects of science and technology. In 2008, her book *Making Scientific Instruments in the Industrial Revolution* (Ashgate, 2007) won the prestigious Paul Bunge Prize.

Simon Naylor is Senior Lecturer in Human Geography at the University of Glasgow. His research interests lie in the histories of science, technology, and exploration. He is the author of *Regionalizing Science: Placing Knowledges in Victorian England* (London, 2010). He is currently investigating the historical geographies of meteorology and of climatology in Britain since 1700.

Eoin Phillips recently completed his PhD in the history of science at the University of Cambridge, UK.

Eugene Rae is Principal Librarian at the Royal Geographical Society (with the Institute of British Geographers). Since 2012 he has been researching the history of the instruments in the Society's archives with a view to creating as complete a record as possible of the instruments' use.

Catherine Souch is Head of the Research and Higher Education Division (RHED) at the Royal Geographical Society (incorporating the Institute of British Geographers), a position she has held since 2006.

Graham Spinardi is Senior Research Fellow, Science and Technology Studies, University of Edinburgh, UK.

Deborah Jean Warner is Curator, Division of Science and Medicine, Smithsonian National Museum of American History, Washington DC, USA.

Claire Warrior is Senior Exhibitions Interpretation Curator at Royal Museums Greenwich. She is also an AHRC CDA award-holder at the Scott Polar Research Institute, University of Cambridge, where she is undertaking a PhD in Polar Studies.

Charles W.J. Withers is Ogilvie Chair of Geography and Professor of Historical Geography at the University of Edinburgh. He is the author, amongst other books, of *Placing the Enlightenment: Thinking Geographically about the Age of Reason* (Chicago, 2007), and, with Innes Keighren and Bill Bell, of *Travels into Print: Exploration, Writing, and Publishing with John Murray, 1773–1859* (Chicago, 2015).

Preface and Acknowledgements

The emergence as a distinct area of interdisciplinary enquiry of what can be termed the geography of science, and, even, the historical geographies of science, has in the last 10 to 15 years or so become a distinguishing feature of research by geographers, historians of science, and others interested in the history of exploration. In a variety of ways, workers in these fields have demonstrated the importance of place and space to the conduct of science, have been attentive to differences in the sites and venues of science's making, and they have illuminated the variable production, circulation, and reception of scientific knowledge as it has been made in, and moved across, space. This work has been paralleled by new studies in the history of exploration, by an evident 'spatial turn' in the humanities and the social sciences, by revitalised and theorised histories of geography, and from new insights in the history of the book. What has been missing, or, at best, much under studied, is critical attention to that major part of science's geographies and to the practices and 'machinery' of geographical and scientific exploration, namely the technology and the instruments that facilitated it. Simply put, while much productive attention has been given to the mapped facts, exploratory personnel, and textual results of geographical and scientific exploration, often in historical context, too little attention has been paid to the instruments used in doing science, to the doing of geography as a form of science, and, more broadly, to questions concerning the technologies of exploration and the technical dimensions to geography and science's undertaking. This book is an attempt to bridge this gap by bringing together scholars from different fields – in geography, history, the history of science, and, importantly, those with professional curatorial and technical expertise – in order to promote conversation and outline the possibilities for new research around the connections between geography, technology, and the instruments of science and exploration.

More particularly, this collection has its origins in successful application to the Small Research Grant scheme of the British Academy, in association with the Royal Geographical Society (with the Institute of British Geographers). The award supported a programme of three one-day research seminars held in December 2011, May 2012, and October 2012. The first and last of the seminars were held in the Institute of Geography in the University of Edinburgh, the second at the Royal Geographical Society in London. One of the key aims was to bring together academics working on the history of exploration, science, and practical geographical enquiry with the formal curatorial expertise of staff with responsibilities for instruments and the history of technology within major national repositories. In total, 18 substantive papers were given, seven of them by museum curators or instrument historians (from the National Maritime Museum, National Museums Scotland, Royal Observatory Greenwich, the RGS-IBG, and the Smithsonian in Washington DC). In the RGS, delegates had the opportunity to see at first hand some of the instruments of exploration held within the collections of the Society, and we acknowledge particularly the work of Eugene Rae and his colleagues for facilitating this. The papers presented in these two cities and three meetings provided the basis for this book, although not all the papers presented in the seminars feature here. In 2013, as the idea for this collection began to take more serious shape, we invited Simon Naylor to participate in the intended collection. We are delighted to incorporate his work as we are for all our contributors. Our contributors make their own acknowledgements in their respective chapters to those persons who have assisted them in their work: here, it is our joint great pleasure to acknowledge not only the quality of their individual chapters, but also their collective engagement with the project and their tolerance of us as we have worked to bring this collection forward for publication. We acknowledge with pleasure the support we have received from Valerie Rose and Katy Crossan and Aimée Feenan and the publishing staff at Ashgate, and the encouragement of Robert Mayhew to bring forward our proposal for formal consideration. We would also like to thank the British Academy for its support; Carolyn Anderson for assistance with the management of the Edinburgh meetings; the Human Geography Research Group of the School of GeoSciences, University of Edinburgh for additional help; and Catherine Souch, Head of the Research and Higher Education Division of the RGS-IBG, for help with the London meeting. It is our hope that in addition to reflecting the stimulating papers that form the basis to this book, the essays here should promote new interdisciplinary connections between those researchers working on the histories of geography and exploration and those on the histories of science and of technology.

Fraser MacDonald and Charles W.J. Withers (Edinburgh)

Introduction: Geography, Technology and Instruments of Exploration

Fraser MacDonald and Charles W.J. Withers

Questions of History and of Method

The instruments and technologies of science and their role in geography and exploration is by no means a new object of historical scholarship. Almost 85 years ago, Eva Taylor, the first woman to hold a chair in British academic geography, published the first of several books charting the connections between mathematical practice, instrumental usage, global exploration, and geographical enquiry.[1] Taylor's chronological range was from the late fifteenth to the mid-nineteenth century; her focus was strongly biographical, and her work overall was motivated by a concern to chart progress in the idea of exactitude, and of its accomplishment, across mathematical geography, navigation, and surveying. Taylor's world was one of craft practices and bespoke makers. She wrote of 'mathematical' and 'philosophical instruments' – as contemporaries termed them – rather than of scientific instruments as mass manufactured precision devices associated with the development of 'modern' science. Method in Taylor's early modern period was essentially concerned with the classification and systematisation of knowledge, rather than with the addition of new material according to agreed empirical procedures.[2]

From the late seventeenth century, however, natural philosophy was distinguished by new advances in the processes of knowledge acquisition. Experimentation required repetition. The developing use of mathematics allowed both the measurement of phenomena and a common language by which it might be transmitted. Information generated through observant travel required an indication of its credibility, via spoken or written testimony from authoritative witnesses.

---

1 Eva G.R. Taylor, *Tudor Geography, 1485–1583* (London: Methuen and Co., 1930); *Late Tudor and Early Stuart Geography 1583–1650* (Cambridge: Cambridge University Press, 1934); *The Mathematical Practitioners of Tudor and Stuart England* (Cambridge: Cambridge University Press, 1954); *The Mathematical Practitioners of Hanoverian England, 1714–1840* (Cambridge: Cambridge University Press, 1966).

2 Robert Mayhew, 'Geography's English Revolutions: Oxford Geography and the War of Ideas, 1600–1660', in *Geography and Revolution* (ed.) David Livingstone and Charles W.J. Withers (Chicago: University of Chicago Press, 2005): 243–72.

Just as observation and recording required writing as a basis to reporting, so the notebook in the field and in the laboratory came to replace memory as a guide to knowledge.[3] 'The rationale for travel note-taking', writes Marie-Noëlle Bourguet, 'derived from the twin dangers of an unruly observation in the field and an unreliable memory'.[4] For natural philosophers, only on-the-spot recording could qualify as the hallmark of accurate observation.

By the late eighteenth century, questions of method in an increasingly disciplined natural philosophy – what from the 1830s would become Science and the sciences – were predicated not only upon the regulation of observation and of inscription, but also upon the epistemic authority of numbers and of measurement. Observation, inscription, reliance upon instrumentation, and trust in numbers went hand in hand.[5] There is, of course, no clean break between early modern method with its focus upon rhetoric, order, and system, and later more strictly scientific method with its appeal to number, procedure and verification as part of emergent disciplinary procedures. Yet by the 1830s, method in science insisted upon trained observation, improved written recording, repetition of numerical measurement, and a reliance upon precision instrumentation. As the mathematician-astronomer John Herschel exhorted in 1830, 'In all cases which admit of numeration or measurement, it is of the utmost consequence to obtain precise numerical statements, whether in the measure of time, space, or quantity of any kind. ... It is the very soul of science; and its attainment affords the criterion, or at least the best, of the truth of theories, and the correctness of experiments'. Herschel further insisted upon the importance 'of improved and constantly improving means of observation, both in instruments adapted for the exact measurement of quantity, and in the general convenience and well-judged adaptation to its purposes, of every description of scientific apparatus'.[6]

---

3   Judy A. Hayden (ed.) *Travel Narratives, the New Science, and Literary Discourse, 1569–1750* (Farnham: Ashgate, 2012); Jaś Elsner and Joan-Pau Rubiés (eds) *Voyages and Visions: Towards A Cultural History of Travel* (London: Reaktion, 1999); Richard Yeo, 'Between Memory and Paperbooks: Baconianism and Natural History in Seventeenth-Century England', *History of Science* 45 (2007): 1–46.

4   Marie-Noëlle Bourguet, 'A Portable World: The Notebooks of European Travellers (Eighteenth to Nineteenth Centuries)', *Intellectual History Review* 20 (2010): 377–400, quote from page 381.

5   David Cahan (ed.) *From Natural Philosophy to the Sciences: Writing the History of Nineteenth-Century Science* (Chicago: University of Chicago Press, 2003); Lorraine Daston, 'The Empire of Observation, 1600–1800', in *Histories of Scientific Observation* (ed.) Lorraine Daston and Elizabeth Lunbeck (Chicago: University of Chicago Press, 2011), 81–113; Theodore Porter, *Trust in Numbers: Pursuit of Objectivity in Science and Public Life* (Princeton: Princeton University Press, 1995); Mary Poovey, *A History of the Modern Fact: Problems of Knowledge in the Sciences of Wealth and Society* (Chicago: University of Chicago Press, 1998).

6   John F.W. Herschel, *A Preliminary Discourse on the Study of Natural Philosophy* (London: Longman and Rees, 1830), 122, 354, respectively.

To draw a correspondence between, on the one hand, instruments in particular and technologies in general with, on the other, questions of method and modernity in science and in geography requires some care. Instruments were known and used in advancing empirical knowledge of the world well before the late eighteenth and nineteenth centuries (as Taylor's work makes clear). As with the term 'method', there is a crucial difference between early modern 'philosophical' or 'mathematical' instruments and modern early 'scientific' instruments, in the sense of the latter as manufactured precision devices capable of extending human capacities. Herschel's concerns over the 'apparatus' of science speak to questions about the longer-term relationship between science and technology and how we treat that relationship's dimensions over time and in different places. As the historian of science Paul Forman has observed in studying the science-technology nexus, modernity has tended to accord science primacy over technology because of the emphasis upon means: 'science was modernity's prime example of progress through reliance upon a proper means, the scientific method'.[7] But examining method in terms of technical capacity – of how, for example, scientific instruments were used and what epistemological claims were made of them – need not suppose the subordination of technology as a means to the end in view.

Post-modernism has accorded considerable significance to a revitalised social and cultural history of technology. What we may think of as an evident 'technological turn' in science studies has certainly transformed our understanding of the relationship between science and technology, and of technical artefacts as having social and political agency.[8] Yet this body of work is relatively recent and there is much that we do not yet know. Perhaps, as others have claimed, we have simply become accustomed to thinking and conceiving of science as an instrumental activity, producing numbers, measurements, and graphs by way of sophisticated devices, either within the space of a laboratory or on field expeditions. We know that instruments, and the data they produce, were designed to travel; in so doing, they provide templates for epistemic authority within science understood as methodised procedures. But we have too commonly overlooked the technical

---

7   Paul Forman, 'The Primacy of Science in Modernity, of Technology in Postmodernity, and of Ideology in the History of Technology', *History and Technology* 23 (2007): 1–152, quote from page 3.

8   On this matter, see for example Trevor J. Pinch and Wiebe E. Bijker, 'The Social Construction of Facts and Artefacts: Or How the Sociology of Science and the Sociology of Technology Might Benefit Each Other', *Social Studies of Science* 14 (1984): 399–441; Steve Woolgar, 'The Turn to Technology in Social Studies of Science', *Science, Technology, & Human Values* 16 (1991): 20–50; Ronald N. Giere, 'Science and Technology Studies: Prospects for an Enlightened Postmodern Synthesis', *Science, Technology, & Human Values* 18 (1993): 102–12; Trevor Pinch, 'The Social Construction of Technology: A Review', in *Technological Change: Methods and Themes in the History of Technology* (ed.) Robert A. Fox (Amsterdam: Harwood, 1996), 17–35; Hans K. Klein and Daniel Lee Kleinman, 'The Social Construction of Technology: Structural Considerations', *Science, Technology, & Human Values* 27 (2002): 28–52.

means by which those 'results' are obtained: 'how was that ability historically achieved? What role did instruments play in establishing the regularity of the new natural order?'[9]

An immediate difficulty involved in answering these and related questions is in knowing exactly what a 'scientific instrument' is. The term was not in common usage before the nineteenth century.[10] It came to be so because of that association with those advances in method, the authority of numbers, and regulation in procedure that we have documented above. The historians of technology van Helden and Hankins in 1994 offer four salient observations on the role of instruments, broadly defined, in the conduct and outcomes of science: instruments, they argue, confer authority; are created for particular audiences (patrons as well as operating scientists); 'act as bridges between natural science and popular culture'; and, 'even in their traditional site, the laboratory, the role of instruments changes when they are used to study living organisms'.[11] At the same time, historians of technology have studied individual devices such as the thermometer, the chronometer, the barometer, and the telescope, either in relation to their manufacture or with respect to the ways in which, in operation, such devices acted to extend human observational capacities in measuring natural phenomena.[12]

In geography, relatively little has been done to address these issues and to build upon Taylor's pioneering enquiries. Geographer Felix Driver has shown how the methodological instructions enshrined in the Royal Geographical Society's instructive guide, *Hints to Travellers*, first produced in 1854 and which ran to

---

9   Marie-Noëlle Bourguet, Christian Licoppe, and H. Otto Sibum, 'Introduction', in *Instruments, Travel and Science: Itineraries of Precision from the Seventeenth to the Twentieth Century* (ed.) Marie-Noëlle Bourguet, Christian Licoppe, and H. Otto Sibum (London: Routledge, 2002), 1–19, quote from page 3.

10   Deborah Jean Warner, 'What is a Scientific Instrument, When Did It Become One, and Why?' *British Journal for the History of Science* 23 (1990): 83–93.

11   Albert van Helden and Thomas L. Hankins, 'Introduction: Instruments in the History of Science', in *Instruments, Osiris* 9 (1994): 1–6, quote from page 5. As they note, 'Sometimes ambiguity is a virtue, and until we have a better understanding of the role of instruments in natural science, we are better off leaving to the term "scientific instrument" its traditional vagueness'. This issue of *Osiris* is a themed issue of 11 papers on instruments organised under the headings of 'Instruments & Authority', 'Instruments & Audience, 'Instruments & Culture', 'Instruments in the Life Sciences'.

12   Jim A. Bennett, *The Divided Circle: A History of Instruments for Astronomy, Navigation and Surveying* (Oxford: Phaidon, 1987); Gerard L'E. Turner, *Nineteenth-Century Scientific Instruments* (Berkeley: University of California Press, 1983); R. Bud and S. Cozzens (eds) *Invisible Connections: Instruments, Institutions and Science* (Bellingham, WA: Society of Photo-Optical Instrumentation Engineers, 1992); Richard Dunn, *The Telescope: A Short History* (London: National Maritime Museum, 2009); W.E.K. Middleton, *The History of the Barometer* (Baltimore, MD: Johns Hopkins University Press, 2003); Hasok Chang, *Inventing Temperature: Measurement and Scientific Progress* (Oxford: Oxford University Press, 2004).

numerous editions, combined instructions about fieldwork, including which instruments were to be utilised and what training was needed to operate them properly.[13] Historian Kapil Raj has shown how travellers involved in survey work in late nineteenth-century Central Asia themselves became instruments (as did their horses), regulating their paces to standard lengths and disguising their use of precision devices: in this case, accuracy was compromised by a need for secrecy.[14] Similarly, the geographer James Rennell had a century earlier recognised the need to calibrate the going rate of different species of camels as instruments of exploration.[15] 'Go-betweens', those vital on the ground mediators between the explorer and the field, were more usually human interlocutors rather than horses or manufactured devices.[16] Yet several recent edited collections on exploration pay little explicit attention to the technologies and instruments involved.[17]

Why there has been relatively little attention to the technologies of exploration from historians of geography and of science is hard to know. One reason may have to do with geographers' greater reliance upon insights from the history of science in exploring geography's historical making, rather than upon the history of technology. Excepting the few studies cited, researchers have hardly glanced at the instrumental devices which explorers employed to obtain the facts of geography. These are the facts which, later and elsewhere, were transformed from in-the-field sightings and readings to become the printed published narratives upon which explorers' personal credibility and reputation came to rest. Paradoxically, it seems that historians of technology have begun to move away from an established focus upon instrument type, instrument makers, and manufacturers' history. As historian of science Liba Taub has observed, recent work re-engaging with instruments has offered different perspectives, from curators, historians of technology, historians

---

13 Felix Driver, 'Scientific Exploration and the Construction of Geographical Knowledge: Hints to Travellers', *Finisterra* 65 (1998): 21–30; *Geography Militant: Cultures of Exploration and Empire* (Oxford: Blackwell, 2001).

14 Kapil Raj, 'When Human Travellers Became Instruments: The Indo-British Exploration of Central Asia in the Nineteenth Century', in *Instruments, Travel and Science* (ed.) Marie-Noëlle Bourguet, Christian Licoppe, and H. Otto Sibum (London: Routledge, 2002), 156–88.

15 James Rennell, 'On the Rate of Travelling, as Performed by Camels; and Its Application, as a Scale, to the Purpose of Geography', *Philosophical Transactions of the Royal Society of London* 81 (1791): 129–45.

16 Simon Schaffer, Lissa Roberts, Kapil Raj, and James Delbourgo (ed.) *The Brokered World: Go-Betweens and Global Intelligence* (Sagamore Beach, MA: Science History Publications, 2009).

17 Simon Naylor and James R. Ryan (ed.) *New Spaces of Exploration: Geographies of Discovery in the Twentieth Century* (London: I.B. Tauris, 2010); Kristian H. Nielsen, Michael Harbsmeier, and Christopher J. Ries (ed.) *Scientists and Scholars in the Field: Studies in the History of Fieldwork and Expeditions* (Aarhus: Aarhus University Press, 2012); Dane Kennedy (ed.) *Reinterpreting Exploration: The West in the World* (Oxford: Oxford University Press, 2014).

of science and others, over how instruments did the work they did, and what happened when they did not. Such inter-disciplinary enquiry, in keeping with the concerns of this volume, treats instruments as objects capable of multiple interpretive reading: as tools; as icons of manufacturing competence; as 'precision' devices whose configuration and use required tolerance, both of error and of one's fellow operatives.[18] Questions of what an instrument *does* rather than what an instrument *is* require us to think about the intimate associations between embodied procedure, authority, accuracy, and disciplinary practice. They also require us, as Davis Baird has it, to think about 'thing knowledge' rather than judge truth claims in the authority of texts alone – to think, that is, of the ways in which 'instrument epistemology' works (even if we do not know exactly how the instrument works) to produce representations of the object under study.[19]

These developments have been supported by studies in material culture and attention to the biographies of scientific objects.[20] They also echo that work advanced by some historians of science towards an integrated social constructivist approach in science studies wherein scientific facts, and the technological artefacts which produced them, should be considered together. In turn, these moves have been animated by concerns to recognise the intrinsically social nature of technical artefacts, notably with respect to Actor Network Theory, and by a felt need for a rapprochement between 'science' and 'technology' within science studies. At the same time, there is a need for a re-appraisal of technology's hitherto subordinate role to science in explanation of intellectual and social developments, even, as Forman has argued, in the making of 'modernity' itself.[21]

---

18   Liba Taub, 'Introduction: Reengaging with Instruments', *Isis* 102 (2011): 689–96: the other articles in this 'Focus' section are Jim Bennett, 'Early Modern Mathematical Instruments', *Isis* 102 (2011): 697–705; Simon Schaffer, 'Easily Cracked: Scientific Instruments in States of Disrepair', *Isis* 102 (2011): 706–17; and Ken Arnold and Thomas Söderquist, 'Medical Instruments in Museums: Immediate Impressions and Historical Meanings', *Isis* 102 (2011): 718–29.

19   Davis Baird, *Thing Knowledge: A Philosophy of Scientific Instruments* (Berkeley and Los Angeles: University of California Press, 2004). The phrase 'Instrument Epistemology' is from the title to the first chapter of this book, pages 1–20.

20   For example, Arjun Appadurai (ed.) *The Social Life of Things: Commodities in Cultural Perspective* (Cambridge: Cambridge University Press, 1986); Lorraine Daston (ed.) *Biographies of Scientific Objects* (Chicago: University of Chicago Press, 2000); *Things That Talk: Object Lessons from Art and Science* (New York: Zone Books, 2004).

21   Bruno Latour, *Science In Action: How To Follow Scientists and Engineers Through Society* (Harvard: Harvard University Press, 1987); Forman, 'The Primacy of Science in Modernity'.

New Possibilities

Here, we consider some of the possibilities for new work that might stem from revitalised inter-disciplinary attention to geography, exploration, science, and technology in historical context. In proposing them, our intention is not to foreclose others; it is, rather, to point to some of the issues with which we framed the series of three seminars on which this collection is based, to identify opportunities for further work, and to lead towards the specific concerns of our contributors.

There is more to know on the relationships between science, technology, and method in terms of individual disciplinary formation. As some contemporary commentators have shown, and as Herschel advocated in the nineteenth century, the rise of method in science depended on instructive guides – how to travel, which instruments to use, and so on – but this issue has not always been studied with reference to specific subject areas as they themselves took shape in the past. Consider the case of geography and, in Britain, of the Royal Geographical Society (RGS), founded in 1830. Early in 1832, the Society outlined its plans to promote geography. Premiums and prizes were to be awarded, including – because nobody quite knew what their subject entailed – the best account of geography's content and its procedures: 'An Essay on the actual state of Geography in all its various departments, distinguishing the known from the unknown, and showing what has been, and what remains to be done, in order to render it an exact Science'. Support was proposed for 'the best mechanical inventions for facilitating the acquisition of Geographical Knowledge', including 'the simplification of instruments'. Chief among the proposals was 'A Traveller's Manual';

> containing a clear and concise enumeration of the objects to which a Geographer's attention should be especially directed; a statement of the readiest means by which the desired information in each branch may be obtained; a list of the best instruments for determining positions, measuring elevations and distances, observing magnetic phenomena, ascertaining temperature and climate, &c; directions for adjusting the instruments, formulae for registering the observations, and rule for working out the results; – adapted to the use, not of the general Traveller alone, but also of him who, in exploring barbarous countries, may be obliged to carry and often conceal his instruments.[22]

These were attempts to guide method and the subject, geography, as a nascent science through the use of text and technology. Julian Jackson's *What to Observe or the Traveller's Remembrancer* (1841), a book which was a formative influence on the RGS's *Hints to Travellers* and which Herschel cited in his 1849 *Manual of Scientific Enquiry*, was insistent concerning instruments. He identified 10 as 'absolutely essential': sextant and artificial horizon; prismatic compass; pocket

---

22 [Anonymous], 'Royal Geographical Society', *Journal of the Royal Geographical Society of London* 2 (1832): v–viii, quote from page vii.

compass; pocket chronometer; portable or mountain barometers; ordinary thermometers; maxima and minima thermometer; barometric, or boiling-point thermometer; hygrometer; clinometer.[23] Portability as well as precision and the capacity for regulated measurement weighed heavily in geography's guidance over instrumentation. Were similar instructions enshrined for the other emergent sciences – of geology, botany, or in meteorology, for example? As individualised sciences became more specialised, what technical developments did practitioners require or come to rely on?[24]

If definitions of 'scientific instrument' have proven problematic, similar difficulties attach to the term 'technologies of exploration'. For historian of science Richard Sorrenson, the ships employed in eighteenth-century voyages of oceanic navigation were akin to floating laboratories, so much so that they might themselves be seen as instruments; of wood and canvas, in their regulation of crew and on-board discipline, and in tolerance of nature's rigours. As many global voyagers discovered, ships could fail and have to be repaired to ensure reliable results, and the secure transmission of information.[25] Is a ship an instrument per se? Not in strictly scientific terms, perhaps, but it is a piece of technology used to scientific ends. If that is so, the technology of exploration might range from rocketry and remote sensing satellites to the pencil to guidance on appropriate clothing. Technology in this sense is not a separate device so much as that suite of objects associated with certain forms of bodily comportment and enhancement of performance in undertaking science. The geologist William Hamilton cautioned the geographical explorer to 'acquire the habit of never quitting his ship without his note-book and pencil and his pocket-compass'.[26] The Arctic traveller Admiral George Back drew from personal experience in offering comments 'On Carrying Chronometers'. Noting how it was 'impossible to avoid jolting pocket chronometers or watches when worn about the person', Back described the elaborate arrangements made of his own clothing to protect the instruments and prevent water and dirt from affecting them.[27]

All this highlights questions relating to practice and performance. As the African explorer Francis Galton was well aware, the quality of instruments was vital:

---

23   Julian R. Jackson, *What To Observe; Or, The Traveller's Remembrancer* (London: John Murray, 1841), 415.

24   Charles W.J. Withers, 'Science, Scientific Instruments and Questions of Method in Nineteenth-Century British Geography', *Transactions of the Institute of British Geographers* 38 (2013): 167–79.

25   On this point, see Simon Schaffer, 'Easily Cracked: Scientific Instruments in States of Disrepair', *Isis* 102 (2011): 706–17.

26   William J. Hamilton, 'Geography', in *A Manual of Scientific Enquiry: Prepared for the Use of His Majesty's Navy: And Adapted for Travellers in General* (ed.) John F.W. Herschel (London: John Murray, 1849), 127–55, quote from page 129.

27   George Back, 'On Carrying Chronometers', in *Hints to Travellers* (ed.) Francis Galton et al. (London: John Murray, 1854), 25.

We are very far indeed from thinking that makers of sextants have yet met all the wants of land travellers, but we *know* that good results can be obtained by such instruments as are to be bought from any good optician. We therefore urge a young explorer to make *these* his mainstay; and if he takes other instruments, to do so more for the purpose of testing and reporting on their performances, than in relying in entire confidence upon them.[28]

John Ross's 1818 Arctic voyage was literally a testing trip: of his officers and crew, and of the scientific instruments on board. His published narrative *Voyage of Discovery* incorporated a report on the effectiveness and reliability of each device. It did so because instruments broke, ran slowly, or could not be easily calibrated (as Back testified). Ross outlined the use made of the seven chronometers carried on board his ship, the *Isabella*, for determining longitude: 'The charge of winding up these chronometers was intrusted to Captain Sabine; And the sentinel at the cabin-door had orders to call him for that purpose at 9 o'clock; and this sentinel could not be relieved by the next, at noon, unless he could report that the chronometers were wound up, (or said to be so) by Captain Sabine'. Sabine and the ship's crew were on the watch in more ways than one. As Ross further reported:

A few days, however, after we had sailed 2,151 [the number refers to an Arnold's chronometer provided by the Admiralty] was unfortunately forgotten to be wound up; and as No. 523, [another Arnold chronometer, in Sabine's own possession] which was worn in Captain Sabine's pocket, altered very much by the effect of heat or cold, it was rejected by me in the calculations for longitude; and no. 2,151 having met with an additional accident in falling out of my hands, was also rejected for the voyage, and the watches were made use of for observing.[29]

This is a story of training the hands, on board ship, to manipulate technology. As naval historian Randolph Cock argues, these are important matters since many of those men who undertook scientific measurement using instruments were not 'scientists' in any strict sense, but naval officers. The emphasis upon practical method over observing, measuring, recording, and reporting was a declaration of methodological simplicity consistent with the training of British naval personnel as themselves regulated instruments, reliable inscription devices suitable for operation in different environments, each capable – like the hand-held precision

---

28   Francis Galton (ed.) *Hints to Travellers*. Second Edition (London: John Murray, 1878), 2–3.

29   John Ross, *A Voyage of Discovery, Made Under the Orders of the Admiralty in His Majesty's Ships* Isabella *and* Alexander, *for the Purpose of Exploring Baffin's Bay and Inquiring into the Possibility of a North-West Passage* (London: John Murray, 1819), cxxvii–cxxxii.

devices they used daily – of tolerating hardship in use whilst maintaining accuracy of record.[30]

To matters of portability and reliability, so we must add dexterity and regularity, of the operator as well as of the instrument. That is why Back, Jackson, Hamilton, and Herschel wrote as they did. If these examples highlight questions of instrumental status, with some devices of the same type being prized more highly than others as a reflection of the manufacturer's status and reliability, they point also to questions of human operational fallibility. As he lay ill during an eventually fatal journey to Timbuctoo in the 1820s, the African explorer Alexander Gordon Laing reported how, during his illness, 'my meteorological observations ceased, and it was with a grief bordering on distraction that I thought of my chronometer, which, as nobody could wind but myself, had unavoidably gone down. I had not been able to take a single observation at Falaba, and had procrastinated from time to time the examination of its rate, which I reason to think was altering'.[31]

Interest in this instance relates not just to the links between numerical observation, accuracy, and bodily frailty. Laing's experience points to that 'epistemic gap' between truth claims about exploratory certainty and evidence which, as the result of technical and human failure, reveals shortfalls in recording, numbering, and, even, in knowing quite where one was at all. Many narratives of exploration document just such matters: yet we have paid too little attention to fallibility, and to how truth claims about science and exploration were made *despite*, not because of, the instruments used.

For the results of exploration and of scientific experimentation to travel, science's technical devices and the things they measure had to be calibrated to effective standards. Studies in the history of metrology have revealed, however, a bewildering array of different units of measurement, varying practices of customary usage, and shown of the metre that attempts to install it as the uniform metrological base unit in measurement of the world foundered for reasons of politics and of geography.[32] For historian of science Simon Schaffer, questions of metrology are an absent presence, constant yet often unacknowledged, and may be seen in several ways: in terms of national difference and of nationalisation; commodification; machinofacture; imperialism; in relation to discipline; and in terms of value. For Schaffer, 'The history of metrology demonstrates that its institutional regulation is also, and precisely, a value system. Metrology's

---

30   Randolph Cock, 'Scientific Servicemen in the Royal Navy and the Professionalisation of Science, 1816–55', in *Science and Beliefs: From Natural Philosophy to Natural Science, 1700–1900* (ed.) David Knight and Matthew Eddy (Aldershot: Ashgate, 2005), 95–111.

31   Alexander Gordon Laing, *Travels in the Timannee, Kooranko, and Soolima Countries, in Western Africa* (London: John Murray, 1825), 88.

32   Witold Kula, *Measures and Men*. Translated by R. Szreter (Princeton: Princeton University Press, 1986); on the metre, see Ken Alder, *The Measure of All Things: The Seven-Year Odyssey that Transformed the World* (London: Little, Brown, 2002).

apparently contradictory demands for institutional insulation and ever wider spatial integration stem from and embody the political and economic conflicts of the modern social order'.[33] In sum, even the units commonly taken for granted in studies of science and exploration, geography and technology are themselves matters of social and political authority.

The Focus of this Collection

The chapters that make up this volume do not all engage in detail with each of the issues discussed above. Rather, we have here brought together contributions that, individually, address and illuminate specific circumstances relating to the workings of geography and of science, technology, and instruments in historical and geographical context and whose purpose, collectively, is to extend further interdisciplinary study of them. The contributions cover a chronological range from the later eighteenth to the second half of the twentieth century, and variously consider hand-held precision devices, texts of technical and moral guidance, rockets, and aircraft amongst the many objects of attention.

Rebekah Higgitt examines the significant role of Nevil Maskelyne, Britain's Astronomer Royal, in promoting later eighteenth-century British voyages of exploration. In establishing as he did rigorous criteria of reliability, functionality and accuracy – in the intending operators as well as of the instruments – Maskelyne helped establish standards as to what the well-equipped expedition should carry, even to merit it being called an 'expedition'. Authoritative standards for the conduct of science at sea did not come easily or cheaply. Eoin Phillips discusses the astronomical and navigational training in instruments given to sailors. His chapter raises questions less of instruments *per se* and more of the presumed technical and the moral competence of users. He also demonstrates the importance of the ship's log book, not as something that should now be read as an accurate record (admitting that ship's logs are a valuable source of data for climate historians), but as something which then provided a crucial guide to the ship's going rate and its position, a form of textual regulation as to when recordings were to be made and of what in particular. Richard Dunn considers the compass as an instrument of navigation with reference to the first polar voyage of John Ross and the nascent science of geomagnetism. Sea trials were just that: experimental events in harsh conditions to judge the accuracy of different devices. Dunn reveals how notions of tolerance and accuracy were always contingent: 'to agree tolerably' as he puts it – men with one another as well as instruments in their readings – was, often, all that could be expected: questions of reliability were, to an extent, judgements upon the manufacturer as well as upon the operative and the device itself. Simon Naylor

---

33 Simon Schaffer, 'Modernity and Metrology', in *Science and Power: The Historical Foundations of Research Policies in Europe* (ed.) L. Guzzetti (Luxembourg: European Communities, 2000), 71–91, quote from page 87.

examines the use and development of meteorological instruments at sea in the early nineteenth century. Taken together, these five chapters highlight questions of standards, of technology understood in terms of the manufacturers' status and instrumental credibility, users' competence and the values associated with 'proper' instrumental usage. In echoing Arnold and Söderqvist's work on medical instruments and museology, they point to the difficulty of writing a 'felt' history, that is, to documenting the learned manipulative skills of instruments' employment.[34]

Four chapters focus upon instruments and technologies of exploration in modern institutional and curatorial context. With reference to the holdings of the National Maritime Museum, John McAleer and Claire Warrior caution about the valorisation of scientific instruments, and the need to see them as part of wider and different resources for knowing about exploration. Alison Morrison-Low, looking at the National Museum of Scotland's instrument holdings, and Eugene Rae, Catherine Souch, and Charles Withers describing the instruments held by the Royal Geographical Society, demonstrate the importance of acquisition policies and of provenance in determining the biography of instruments and in illuminating historical and geographical associations between devices, explorers, and scientific claims. In Edinburgh's case, the focus was on marine science.[35] In London in the RGS, written records survive to show instruments at work throughout the world. The practical efficacy of some devices was not echoed in the moral economy of their operators: several instruments were borrowed and never returned, others broke and were not repaired or replaced. For Deborah Warner, the 'performance characteristics' of geodetic instruments were all important, and did not develop in isolation. The technical capacities to measure the 'trembling earth' involved university staff in different countries. A global metrology for seismology emerged only gradually.

Our final suite of chapters has as its collective focus the geographical study of the earth from above. Peter Collier examines the military and technical bases to aerial photography and the presumption of operating authorities that technical developments necessarily led to completeness of coverage. As Isla Forsyth shows in exploring the connections between camouflage and military surveillance in World War II, instruments whose presumption was ever closer inspection and greater accuracy could be thwarted by what were, on the surface, mundane technologies of earthly disguise. Fraser MacDonald takes up the story of Frank Malina, rocket scientist, and the problems associated with getting instruments to leave the earth's orbit and, once left, return in one piece, while delivering credible results en route. No less than for eighteenth-century oceanic voyages of terrestrial exploration, the path of rocketry's rise to extra-terrestrial exploration was strewn with obstacles and questions of tolerance at once political, epistemological, and technological. In

---

34   Ken Arnold and Thomas Söderqvist, 'Medical Instruments in Museums: Immediate Impressions and Historical Meanings', *Isis* 102 (2011): 718–29.

35   In the seminar in question, the paper was delivered by Dr Tayce Phillipson of the National Museum of Scotland: we are pleased to acknowledge her contribution.

his essay on the UK's Linesman radar early warning system in the 1960s, Graham Spinardi's account is likewise of the complex workings – and failure – of a socio-technical system whose accuracy was finally measured in terms of political value.

These chapters point to the ways in which instruments construct the objects of their own exploration, making the field and the subject through uneven passage over space. They additionally demonstrate the military context behind many technologies of exploration and the associations in the twentieth century between accuracy, surveillance, long-distance telemetry, and cold war geopolitics.[36] Readers will be aware that such issues, as for those discussed in other chapters, speak in the twenty-first century to the continued importance of understanding the many and differently configured connections between science, technology, geography, and exploration today, in the future, and in the past.

---

36 On these issues, see Matthew Farish, *The Contours of America's Cold War* (Minneapolis: University of Minnesota Press, 2010), and the essays making up Simone Turchetti and Peder Roberts (eds) *The Surveillance Imperative: Geosciences during the Cold War and Beyond* (Palgrave Macmillan: London, 2014), especially the introductory essay by Simone Turchetti and Peder Roberts, 'Knowing the Enemy, Knowing the Earth', 1–19.

Chapter 1

Equipping Expeditionary Astronomers: Nevil Maskelyne and the Development of 'Precision Exploration'

Rebekah Higgitt

In examining the contribution of Nevil Maskelyne (1732–1811), Britain's fifth Astronomer Royal, to voyages of scientific exploration in the second half of the eighteenth century, this chapter has three aims. First, it will explore how Maskelyne came to undertake an advisory role for expeditions on behalf of the Royal Society, the Board of Longitude, and the Admiralty. Partly as a result of his prior experience, he was more active in this area than his predecessors and helped to develop a standardised approach through selecting instruments and personnel, writing instructions, and defining objectives. This ensured that naval voyages of exploration and survey could make physical observations of the greatest possible precision and so become part of wider moves toward quantification and precision measurement in overcoming problems of geographical distance and individual difference.[1] Secondly, the chapter suggests that Maskelyne's role in the development of the British tradition of scientific exploration has been under appreciated. Thirdly, the chapter raises the question of what should be understood by 'instruments of exploration' and the extent to which the lists of instruments that Maskelyne drew up can be seen as definitive. In considering this third aim, one of the pertinent issues is the expected level of precision. The expeditions with which Maskelyne was most closely involved were ones aiming at high levels of precision and generally relied upon ships to transport the required mass of instrumentation. While always vulnerable to loss or damage, numerous instruments and reference works of considerable size, weight and delicacy could nevertheless be carried overseas.[2]

The 'precision exploration' discussed here questions our definition of 'exploration' as well as of 'scientific instrument'. The instruments used on expeditions within the British Isles could be identical to those taken to the South Seas, while the

---

1   Marie-Noëlle Bourguet, Christian Licoppe and H. Otto Sibum, 'Introduction', in Marie-Noëlle Bourguet, Christian Licoppe, H. Otto Sibum (eds) *Instruments, Travel and Science: Itineraries of Precision from the Seventeenth to the Twentieth Century* (London: Routledge, 2002), 1–19.

2   Richard Sorrensen, 'The Ship as a Scientific Instrument in the Eighteenth Century', *Osiris* (2nd Series), 11, *Science in the Field* (1996): 221–36; Simon Schaffer, 'Easily Cracked: Scientific Instruments in States of Disrepair', *Isis* 102 (2011): 706–17.

navigational tools essential to James Cook's circumnavigations were also used to survey known locations. Different levels of precision might be required on different voyages, or for different purposes within one voyage. While fixing a position on land required repeated observation with the best instruments, coastal surveys built on these fixes might rely on chains, rods and observation from boats. Astronomy, survey, hydrography, navigation, and meteorology were all performed on such voyages, but each required a different approach, as did exploring the interior of a landmass. If, however, high levels of precision were desired, good instruments were required, their quality affirmed by the names of their makers, who were often identified early in the planning process. Precision instruments were expensive and so were put to repeated use. Some were extremely well travelled, being made available from the collections of the Royal Society and the Board of Longitude, as well as – increasingly – the Royal Navy.[3] The use of particular makers and identical specifications were means by which the reliability of observations could be, if not guaranteed, at least a reasonable ambition, especially if supported by reliable observers, accurate tables, and good recording systems.

Instruments of precision observation form the core of this chapter. They, and their associated records of observations, were Maskelyne's principal concern, unlike the jars, nets or cases of natural history and the logs, journals, and scrapbooks that also formed part of the material culture of expeditions and voyages.[4] Whether the observers themselves might be considered 'instruments of exploration' is moot, but, as the chapter demonstrates, Maskelyne played a vital role in selecting and training as well as equipping and instructing such individuals. Their reliability was as important – and as hard to maintain – as that of the instruments. These men were, nevertheless, the means by which the practices valued by Maskelyne were embedded in Royal Naval survey voyages. His alignments of instruments, projects, observers and the interests of the Royal Society and Navy helped shape the nineteenth-century role of the scientific serviceman.[5]

The collection of physical data had always had a basic purpose in navigation. It added to knowledge about the world's navigable routes and potentially exploitable

---

3   Clocks in particular had long and active lives: see Derek Howse and Beresford Hutchinson, 'The Saga of the Shelton Clocks', *Antiquarian Horology* 6 (1969): 281–98.

4   On logbooks and scrapbooks as part of naval practice at a slightly later date, see Felix Driver and Luciana Martins, 'John Septimus Roe and the Art of Navigation, *c.*1815–1830', *History Workshop Journal* 54 (2002): 144–61, and, more generally, Marie-Noëlle Bourguet, 'A Portable World: The Notebooks of European Travellers (Eighteenth to Nineteenth Centuries)', *Intellectual History Review*, 20 (2010): 377–400.

5   'Scientific servicemen' are discussed in David Philip Miller, 'The Royal Society of London 1800–1835: A Study in the Cultural Politics of Scientific Organization' (PhD Dissertation, University of Pennsylvania, 1981), 120–32. Miller refers to 'an inner core of naval officers many trained initially on Cook's great voyages and later associated with the Admiralty's Hydrographic Office' and also, increasingly, individuals trained at the Royal Naval College at Portsmouth, who 'formed an inner core of 'scientific' naval officers': pages 120, 122 respectively.

lands. With the introduction of new instruments and techniques (from the 1760s especially, marine timekeepers, sextants and astronomical tables), recording data was also necessary to test them and make recommendations about their use.[6] Making and recording observations of land and the skies served the purpose of fixing positions and creating improved charts. The investigation of geomagnetism, in order to understand the behaviour of compasses, likewise aided navigation. Hydrographic investigation included the essential business of sounding, to check or map the depth of the sea floor, and the observation of tides, weather and currents. Other matters such as air pressure, temperature, saltiness and the specific gravity of water were also recorded. Different instruments were required for these observations, although some – portable telescopes, octants, compasses and watches, for example – had a range of uses.

Collecting this information also served more obviously scientific projects. The rhetoric of science improving navigation and of exploration aiding science worked both ways. Rather than precision being demanded by the state, justifications and appeals for support presented science and exploration as interlinked and inevitably improving trade and national prestige. This allowed 'the rhetoric of precision', as Norton Wise put it, to 'acquire the power to carry conviction'.[7] Measurement of variations in magnetism and gravity fed into theories about the Earth as well as into mapping its features. Gravimetric investigations undertaken to investigate the question of the shape of the Earth had practical implications for charting and for calibrating one of the instruments – the pendulum clock – used to define the most accurately known locations. Geodetic survey had, simultaneously, clear political and economic rationale: measuring and comparing degrees of latitude helped to establish the figure of the Earth. Pierre Louis Maupertuis justified this project in the 1730s both as a matter of 'Curiosity' and 'Speculation' among philosophers and in typically down-to-earth fashion: 'if the distances of Places are not very well known, to what dangers must the Ships be exposed that are bound for them!'[8] Similarly, observations of the transits of Venus might be seen as of purely astronomical interest but the utility of an accurate scale for the solar system in improving astronomical tables meant that this project too was linked to navigation. The interlinked nature of these projects, and the means by which they could be pursued, was exploited fully by Maskelyne and was key to their

---

6   See the first part of M. Norton Wise (ed.) *The Values of Precision* (Princeton: Princeton University Press, 1995) on the 'Enlightenment Origins' of precision measurement and precision instruments, beginning with astronomy and extending to the allied practices of survey and navigation.

7   Wise, *Values of Precision*, 92. Drawing on John Heilbron and Ian Hacking, Wise suggests that quantification was 'derived from the needs of administrators for reliable information' rather than 'as a campaign of mathematicians'; Wise, *Values of Precision*, 5.

8   Pierre Louis Maupertuis, *The Figure of the Earth, Determined from Observations Made by Order of the French King, at the Polar Circle* (London: T. Cox, 1738), 3–4.

inclusion within the remit of the Board of Longitude by the last quarter of the eighteenth century.

Maskelyne and Expeditionary Observation

Maskelyne was Astronomer Royal between 1765 and 1811, a period coincident with the first flourishing of British voyages of scientific exploration and survey.[9] We may think in this context, *inter alia*, of Cook's circumnavigations (1768–71, 1772–75, 1776–79); the attempt on the North Pole by Constantine Phipps (1773); George Vancouver's expedition to the north-west Pacific (1791–95); and Matthew Flinders' circumnavigation of Australia (1801–1803). Maskelyne's position connected him with the key institutions involved in these projects. The Royal Observatory, Greenwich, (ROG) was linked, financially, to the Board of Ordnance and, in terms of its purpose of improving celestial navigation, to the Admiralty. As Astronomer Royal, Maskelyne was *ex officio* a Commissioner of Longitude, and he was to become the most significant figure on the Admiralty-financed Board of Longitude. He was also a key individual within the Royal Society, often on its Council and its chief correspondent on astronomical matters. His education, career, and family also underscored connections within the astronomical, instrument-making, and maritime communities.

Derek Howse's biography is clear about Maskelyne's involvement in the expeditions named above and others, including the British transit of Venus expeditions of 1761 and 1769 and the surveying of the Mason-Dixon Line (1763–67), in addition to his expeditions to St Helena (for the 1761 transit of Venus) and to measure the effects of gravity at Schiehallion in Perthshire (1774). However, more remains to be said about how Maskelyne developed his role, moving between these different institutional settings, and how the scientific arrangements for which he was responsible were made and standardised. Perhaps surprisingly, in books on scientific voyaging, Maskelyne is often left out of the picture. Joseph Banks usually gets the credit of being the link between the worlds of science and government-funded exploration.[10] While much of the attention regarding the exploration of new lands focused on botany, zoology and ethnography, Maskelyne was involved with the allied but, in terms of equipment and personnel, separate projects of collecting physical data. His role is usually

---

9   See Derek Howse, *Nevil Maskelyne: The Seaman's Astronomer* (Cambridge: Cambridge University Press, 1989); Rebekah Higgitt (ed.) *Maskelyne: Astronomer Royal* (London: Robert Hale, 2014).

10   Gascoigne, *Science in the Service of Empire*. Andrew S. Cook considers the Royal Society and exploration before Banks's presidency in 'James Cook and the Royal Society', in Glyndwr Williams (ed.) *Captain Cook: Explorations and Reassessments* (Woodbridge: Boydell Press, 2004), 37–55, but misses the significance of Maskelyne's experience and role.

revealed in only piecemeal fashion, cropping up only then to slip away.[11] He was described as a 'small invisible man', and it is possible that, even in his own day, his contribution was less widely recognised than it might have been if he had the kind of dominant character Banks enjoyed.[12]

The emphasis on collecting physical data on voyages and Maskelyne's involvement as Astronomer Royal may look inevitable; yet such seeming inevitability was the result of individuals' actions. The position of Astronomer Royal and the institution of the ROG could have developed otherwise than they did: Maskelyne's interests and actions were crucial in creating a lasting tradition for the use of precision instruments on board naval vessels. Of his predecessors, only Edmond Halley had taken a similar practical interest in voyaging. John Flamsteed, James Bradley, and Nathaniel Bliss (and his successor John Pond), while attending to the ROG's foundational purpose of helping solve the problem of finding longitude at sea, each concentrated on basic positional astronomy. They had focused neither on making their data widely available nor on close involvement with maritime exploration, beyond an advisory role in relation to trials of instruments associated with the Board of Longitude.[13]

Bradley was certainly involved in this work for the Board and the Royal Society, although in one crucial case – the planning of expeditions by the Royal Society for the first eighteenth-century transit of Venus – he largely delegated the work. He and the Society were only prompted to action after learning that the French planned expeditions, and the British expeditions were somewhat hastily assembled in response. Thus on 5 June 1760 the Royal Society heard read a printed memoir, dated 27 April, by Joseph-Nicolas Delisle regarding the June 1761 transit. Action was urged by a motion of 19 June and Bradley was asked to suggest people who could undertake an expedition and to provide advice and instructions.[14] There and then Bradley gave a provisional list of instruments that would be required:

---

11   Maskelyne appears just four times in the index of Margarette Lincoln (ed.) *Science and Exploration in the Pacific: European Voyages to the Southern Oceans in the Eighteenth Century* (Woodbridge: Boydell & Brewer, 2001), twice in Robin Fisher and Hugh Johnston (eds) *Maps to Metaphors: The Pacific World of George Vancouver* (Vancouver: UBC Press, 1993), and once in John Gascoigne, *Science in the Service of Empire: Joseph Banks, the British State and the Uses of Science in the Age of Revolution* (Cambridge: Cambridge University Press, 1998). Maskelyne's activities are often hidden within mention of the Royal Society or Board of Longitude as collective identities, or, as in Sorrensen (fn. 2 above), conflated with that of ship's commanders and the general business of navigation and survey.

12   Johan Henrik Lidén called him 'en liten osynlig man', letter to Fredrik Mallet, 10 July 1769, quoted in Andrea Wulf, *Chasing Venus: The Race to Measure the Heavens* (London: William Heinemann, 2012), 124.

13   Eric G. Forbes, *Greenwich Observatory: The Royal Observatory at Greenwich and Herstmonceux 1675–1975, vol. 1: Origins and Early History 1675–1835* (London: Taylor & Francis, 1975).

14   Howse, *Nevil Maskelyne*, 21–2.

'Reflecting Telescope of Two Foot with Dolland's [sic] Micrometer, Mr. Dolland's Refracting Telescope of Ten Feet; a Quadrant of the radius of Eight Inches; and a Clock or time piece'. Harry Woolf rightly noted that 'to some extent Bradley's list can be regarded as the minimum requirements for any astronomical expedition'.[15] Any expedition could become an astronomical one using this equipment, if precise determination of positions on land was desired. These instruments, by respected makers like Dollond, appear time and again in subsequent lists.

This list was merely preliminary, however, and when Bradley presented a more considered list to the Society on 3 July 1760, it had been drawn up by Maskelyne, then a 28-year-old Fellow of Trinity College, Cambridge. Maskelyne had been a Fellow of the Royal Society since 1758, when Bradley, among others, signed his election certificate.[16] It is clear that by 1760 Maskelyne was involved in discussions and recognised the expeditions as a chance to advance his career. On 26 June 1760, between presentation of the preliminary and full lists of instruments, he read a paper proposing that whomsoever was sent to observe the transit on the island of St Helena (a destination only proposed to and accepted by the Council that same day) should also take the opportunity to attempt observations of the parallax of Sirius.[17] This paper put Maskelyne firmly in the frame to be selected as one of the expedition astronomers. It is worthy of note that the proposal was to use the ships and territories of the East India Company, which confirmed its support for the venture by 3 July. Maskelyne had family links with this institution through his brother-in-law Robert Clive, who returned from India that month.[18] Maskelyne may also have assisted John Michell with writing instructions for observing the transit, which were sent to the Presidencies of the East India Company 'at the Sollicitation of some members of the University of Cambridge'.[19]

Maskelyne's paper included details of the principal instrument that would be required for the attempt to measure stellar parallax: a zenith sector. In this he drew on Bradley's and Maupertuis' experiences of using such instruments, at home and overseas, and observations of Sirius made by Nicolas Louis de Lacaille at the Cape of Good Hope.[20] When Bradley presented Maskelyne's list of instruments

---

15   Harry Woolf, 'British Preparations for Observing the Transit of Venus of 1761', *The William and Mary Quarterly* 13 (1956): 499–518, quote from page 504.

16   It seems the two were already acquainted, and Maskelyne had possibly already assisted the Astronomer Royal: Maskelyne's Election Certificate, 27 April 1758, Royal Society, EC/1758/04, shows that the signatories made their recommendation from 'our own personal knowledge'; Howse, *Nevil Maskelyne*, 15.

17   Nevil Maskelyne, 'A Proposal for discovering the Annual Parallax of Sirius', *Philosophical Transactions of the Royal Society of London* 51 (1760): 889–95.

18   Patrick Turnbull, *Clive of India* (Folkestone: Bailey and Swinfen, 1975), 110, notes Clive reached England in July 1760.

19   Howse, *Nevil Maskelyne*, 21.

20   Maskelyne, 'A Proposal for discovering the Annual Parallax of Sirius'. See Nicky Reeves, 'Constructing an Instrument: Nevil Maskelyne and the Zenith Sector, 1760–1774' (PhD thesis, University of Cambridge, 2008), 49–53.

for the transit observations to the Royal Society on 3 July, it included a 10-foot radius zenith sector. This was in addition to three reflecting telescopes (two for St Helena and one for Greenwich), two micrometers, a clock with a compound pendulum, and an 18-inch astronomical quadrant for correcting the clock against the Sun and stars. Bradley explained that these were not available for hire and so must be purchased: £185 for the transit instruments, £100 for the zenith sector. A total of £685 was given for the costs of instruments, travel and expenses for the year on St Helena required in order to measure annual parallax. The Royal Society made an appeal to the King, upping the sum requested to £800. This was granted and, shortly afterwards, an additional £800 grant was made for a second expedition to Bencoolen in Sumatra. Maskelyne was appointed to lead the St Helena expedition on 14 July 1760, and was charged with ordering instruments for both expeditions. The telescopes were made by Dollond and James Short, the clocks by John Shelton and John Ellicott, and quadrants were lent by Bradley and the Earl of Macclesfield.[21]

The voyage to St Helena was the fulcrum of Maskelyne's career. While he failed to observe either the transit (due to clouds) or stellar parallax (due to the limits of eighteenth-century instruments), the voyage was a success inasmuch as he, and his assistant Robert Waddington, used the lunar-distance method of longitude determination on board ship. This led directly to his developing that method and creating, first, the *British Mariner's Guide* (1763) and, second, having secured the role of Astronomer Royal, the *Nautical Almanac* from 1767.[22] The voyage was an opportunity for him to become familiar with navigational and astronomical instruments – sextant, watches, astronomical regulator, reflecting telescope with micrometer, quadrant, thermometers and barometer, as well as the zenith sector – in new settings. As Nicky Reeves has shown, it afforded Maskelyne an opportunity to manage the processes, literally and figuratively, of stabilising instrumentation in order to produce reliable knowledge at a distance.[23] Just as importantly, it put him in an organisational role. He had selected instruments for both expeditions, while the management of his own projects at St Helena, with assistance from Waddington and then Charles Mason (a former assistant to Bradley at Greenwich) gave him experience of physical data collection. As well as solar and stellar parallax, Maskelyne made observations to establish exact position and the going of the clock, both being essential to the accuracy of other observations,

---

21 Howse, Nevil Maskelyne, 23–6.

22 On Waddington's career see Jim Bennett, '"The Rev. Mr. Nevil Maskelyne, F.R.S. and myself": The story of Robert Waddington', in Higgitt, *Maskelyne*, 59–88. It contrasts with Maskelyne's career, suggesting the importance of the latter's Westminster and Cambridge education, personal acquaintance with astronomers and instrument makers, and activity within the Royal Society in enhancing his credibility.

23 Reeves, 'Constructing an Instrument', 57–77; Nicky Reeves, '"To demonstrate the exactness of the instrument": Mountainside Trials of Precision in Scotland, 1774', *Science in Context* 22 (2009): 323–40.

in addition to investigating into local gravity and the figure of the Earth.[24] He additionally claimed to make magnetic and tidal observations and although there is no record of the former having been carried out, Maskelyne and Mason observed a tide pole at James's Fort over several weeks. Comparative observations were a check against each other and accuracy was maximised through repeated observations, generally '40 to 50' and 'sometimes more than 100', to produce median heights.[25]

Maskelyne's St Helena experience, as well as his connection to Bradley, led to his next, and final, scientific voyage overseas. This was to Barbados in 1763–4 on behalf of the Board of Longitude, of which he was not yet a member. He was charged with making observations that would test three possible longitude methods: John Harrison's sea watch (today known as H4), Tobias Mayer's lunar tables, which underpinned Maskelyne's successful use of the lunar distance method, and Christopher Irwin's marine chair, intended to assist in observing Jupiter's satellites at sea.[26] On this occasion, Maskelyne was instructed by the Board's astronomers (Bradley and the Oxbridge professors of astronomy and mathematics), who supplied the most important instruments, including a clock borrowed from the Royal Society. In the minutes of the Board Maskelyne appeared as a hired hand, although he probably became increasingly involved in the arrangements: he had his own instruments to contribute, with an octant and reflecting telescope by John Bird and a watch by John Ellicott.[27] This expedition again involved position-determining observations at sea and in a temporary observatory on land, using octants, sextants, watches, telescopes and clocks. As for St Helena, Maskelyne's notebooks reveal how much adjusting, compensating, and gathering of circumstantial details went into the production of series of apparently routine

---

24   Nevil Maskelyne, 'Observations on a Clock of Mr John Shelton, made at St Helena, in a Letter to the Right Honourable Lord Charles Cavendish, Vice-President of the Royal Society', *Philosophical Transactions of the Royal Society of London* 52 (1762): 434–43. The observations were made more valuable by Mason bringing the clock used for the other transit of Venus expedition (which reached the Cape of Good Hope rather than Bencoolen). Comparison of the effects of gravity at Greenwich, St Helena and the Cape could thus be attempted.

25   Nevil Maskelyne, 'Observations of the Tides in the Island of St Helena: in a Letter from the Rev. Nevil Maskelyne, A.M. F.R.S. to Thomas Birch, D.D. Secretary to the Royal Society', *Philosophical Transactions of the Royal Society of London* 52 (1762): 586–606, quote from page 588.

26   Jim Bennett, 'The Travels and Trials of Mr Harrison's Timekeeper', in Bourguet, Licoppe, Sibum (eds) *Instruments, Travel and Science*, 75–95.

27   Confirmed Minutes of the Board of Longitude, 4 August 1763, Royal Greenwich Observatory Archives, RGO 14/5, p. 49 (http://cudl.lib.cam.ac.uk/view/MS-RGO-00014-00005/53), Cambridge University Library; Howse, *Nevil Maskelyne*, 49.

observations, especially when your observatory had to be established from scratch or was actually moving.[28]

All this experience, considered retrospectively, seems essential to achieving the position of Astronomer Royal, which Maskelyne did after Bliss died in 1764. However, Bradley and Bliss had been university professors, not expeditionary astronomers, as also was Joseph Betts, a possible alternative to Maskelyne in 1764.[29] Rather, what seems to have been essential was Maskelyne's being known personally – as a result of his work for the Royal Society and the Board of Longitude – to key figures such as Lord Macclesfield (President of the Royal Society until March 1764), Lord Morton (the new President), Lord Sandwich (Secretary of State and, previously, First Lord of the Admiralty – and thus a Commissioner of Longitude – in April–September 1763) and George Grenville (Prime Minister, and the previous First Lord of the Admiralty). His family's political connections were important. Morton's parliamentary association had been with the Duke of Newcastle, to whom Maskelyne's father had been a clerk and who assisted the family after the father's death. Maskelyne's sister, Lady Clive, had informed him of the position's availability and assured him that there was support should he apply. Lord Clive later wrote to Grenville, noting 'I must not forget to express how thankful I am for the assistance you have given Mr Nevil Maskelyne to obtain the Regius Professorship'.[30]

Once Astronomer Royal, Maskelyne was immediately involved as Commissioner of Longitude in analysing and acting on the results of the Barbados trial. Within the Royal Society he was quickly co-opted to Council to help deal with the proposal from Charles Mason and Jeremiah Dixon (who had been observers on the other transit of Venus expedition) to extend their survey of the boundary between Pennsylvania and Maryland to measure a whole degree of latitude in order to contribute to attempts to establish the figure of the Earth by Maupertuis, Lacaille, and others. Maskelyne was directed to deal with the additional instructions and instruments that would be required. Instruments included the same Shelton clock Maskelyne had used on St Helena and Barbados, as well as surveying tools like fir measuring rods – made as close to precision instruments as possible by the inclusion

---

28   See, for example, the 'Corrections of Adjustment of Sextant' in his private notebook, 'Observations', Royal Greenwich Observatory Archives, RGO 4/1, Cambridge University Library. Megan Barford's useful introduction to the digitised version of this notebook (http://cudl.lib.cam.ac.uk/view/MS-RGO-00004-00001/53), mentions that it reveals 'the importance of knowing one's hardware' and 'routine shipboard activity' to Maskelyne's work.

29   Howse, *Nevil Maskelyne*, 54.

30   Maskelyne to Anthony Shepherd, 15 October 1764, REG00009/37 (http://cudl.lib.cam.ac.uk/view/MS-REG-00009-00037/352). Howse, Letter from Clive to Grenville, dated 30 September 1765 from Calcutta, in John Malcolm, *The Life of Robert, Lord Clive: Collected from the Family Papers Communicated by the Earl of Powis*, 3 volumes (London: John Murray, 1836): 2, 372; Howse, *Nevil Maskelyne*, 53.

of a brass standard, spirit level, and thermometers.[31] These added to the equipment already in Pennsylvania, including a transit instrument and zenith sector made by John Bird, the latter benefitting from improvements Maskelyne suggested after experiments with his sector.[32] This was already a survey of extraordinary precision for a simple boundary dispute, although Maskelyne further raised the exactness demanded as the survey turned into a scientific expedition focusing on the figure of the Earth. Mason and Dixon were instructed to 'measure the lines carefully over-again' with the new instruments: the fir rods 'were to be compared frequently, and the difference noted, and also the height of the thermometer at the time; for the lines had been all measured before with a standard chain, which, though sufficient for the common purposes of surveying, was by no means to be depended upon in so nice an operation as that of measuring a degree of latitude'.[33]

Placing Precision Observation on Voyages of Exploration

From the later 1760s, Maskelyne, through the Royal Society, Board of Longitude and Admiralty, was involved in the organisation of a significant series of voyages of scientific exploration, especially with preparation for the next transit of Venus from 1767. He was the most influential figure for the British expeditions, choosing observers, settling locations and drawing up instrument lists. The project had a clear astronomical objective and so it was unsurprising that the Astronomer Royal was the key figure in its organisation – after all, it was Bradley to whom the Royal Society had initially turned in 1760. Unlike Bradley, however, Maskelyne was also a veteran – as observer and organiser – of the previous transit. The instruments that he had commissioned for the 1761 transit could be reused, supplemented with new ones, as similar as possible, plus wooden portable observatories. He again authored observing instructions, published as an appendix to the 1769 edition of the *Nautical Almanac*.[34]

As with Maskelyne's St Helena voyage, the opportunity for travel was made to work in several ways. Fixing longitudes was an objective in itself as well

---

31   Howse, *Nevil Maskelyne*, 109; Howse and Hutchinson, 'The Saga of the Shelton Clocks'.

32   Reeves, 'Constructing an Instrument', 87.

33   Nevil Maskelyne, 'Introduction to the following Observations, made by Messrs, Charles Mason and Jeremiah Dixon, for Determining the Length of a Degree of Latitude, in the Province of Maryland and Pennsylvania, in North America', *Philosophical Transactions of the Royal Society of London* 63 (1768): 270–73, quote from page 272.

34   On the preparations and expeditions for the 1769 transit of Venus see Harry Woolf, *The Transits of Venus: A Study of Eighteenth Century Science* (Princeton, NJ: Princeton University Press, 1959), J.C. Beaglehole, *The Life of Captain Cook* (Stanford: Stanford University Press, 1974); Derek Howse, 'The Principal Scientific Instruments Taken on Captain Cook's Voyages of Exploration, 1768–80', *Mariner's Mirror* 65 (1979): 119–35; Wulf, *Chasing Venus*.

as being an essential part of observing the transit. Setting-up a clock to work reliably for astronomical observations also allowed comparison of the length of pendulums at different locations, and so permitted exploration of the effects of gravity locally, and of the figure of the Earth. All this meant that Maskelyne found reason for the Board of Longitude and the Royal Society as well to take an interest, pointing out that 'as the Observation of the Transit itself affords one of the best methods of determining the Longitudes of places', they would profit by sponsoring an additional expedition to Cornwall: 'it would be useful to Navigation to have accurate Observations made at the Lizard Point in order to settle the just Latitude & Longitude thereof which has never yet been done'. The Board agreed, resolving that 'M^r. Bird be directed to make an Astronomical Quadrant of One foot Radius, and Mr Shelton an Astronomical Clock for the purpose of making those Observations'.[35]

The voyages were also an opportunity to continue trials of the lunar-distance method, particularly with the long-distance travel of Cook's expedition to Tahiti and beyond. Cook and the Royal Society-nominated astronomer Charles Green (a former assistant at Greenwich who had accompanied Maskelyne to Barbados) not only observed the transit but also made many lunar-distance observations, being supplied with sextants, the available editions of the *Nautical Almanac*, for the years 1768 and 1769, and two copies of Mayer's lunar tables should they remain overseas beyond 1769.[36] Green also taught officers the processes of observation and calculation, as had Maskelyne and Waddington on the St Helena voyages.[37] With a trained astronomer, naturalist, and artist on board, the expectation was for the gathering of as much data and information as possible; on people, landscapes, flora, fauna, navigation, astronomy, surveying, map-making, geomagnetism, meteorology, and hydrography. The precision instruments were the business of the Royal Society and Maskelyne. The natural historical equipment was largely supplied by Joseph Banks as a private passenger, other equipment by the Admiralty.

The success of Cook's voyage led the Admiralty to mount another within months of his return. This would not, of course, involve a transit of Venus and so it was necessary for Maskelyne to take the initiative to ensure that precision instruments and trained observers would again be included. In October 1771, he

---

35   Confirmed Minutes of the Board of Longitude, 12 November 1768, Papers of the Board of Longitude RGO14/5, p. 175 (http://cudl.lib.cam.ac.uk/view/MS-RGO-00014-00005/179), Cambridge University Library.

36   Howse, 'The Principal Scientific Instruments', 133–4.

37   Cook reported that 'by [Green's] Instructions several of the Petty officers can make and Calculate these observations almost as well as himself: it is only by such means that this method of finding the Longitude at Sea can be put into universal practice': J.C. Beaglehole (ed.) *The Journals of Captain James Cook on his Voyages of Discovery: The Voyage of the Endeavour 1768–1771* (Cambridge: Cambridge University Press, 1955), 392. Maskelyne and Waddington's involvement of officers in lunar distance observations on the voyage to St Helena was described by Jim Bennett in a paper at the 'Longitude Examined' conference, National Maritime Museum, 25 July 2014.

wrote to Lord Sandwich, then again first Lord of the Admiralty, hoping that the new voyage 'may be rendered more serviceable to the improvement of Geography & Navigation than it can otherwise be if the ship is furnished with Astronomical Instruments as this Board hath the disposal of or can obtain the use of from the Royal Society and also some of the Longitude Watches; and, above all, if a proper person could be sent out to make use of those Instruments & teach the Officers on board the ship the method of finding the Longitude'.[38]

The primary aims, from the point of view of Maskelyne and the Board of Longitude, which approved the idea on 28 November, were: to test timekeepers made by Larcum Kendall and John Arnold; to continue trialling and teaching the use of the *Nautical Almanac*; and to make accurate determinations of longitude on land, both as a means of checking the accuracy of the lunar-distance observations and timekeepers and to assist accurate charting. Once observers were included, other observations could easily be added to their instructions. The Board resolved: 'That the Astronomer Royal & the rest of the Professors present be desired to look out for two persons properly qualified & willing to go upon the above service ... And, That they be further desired to prepare a Draft of Instructions proper for the said Persons and also a List of the necessary Instruments'.[39] The observers chosen were William Bayly, a former assistant at the ROG and observer for the 1769 transit of Venus expedition to Norway, and William Wales, a computer for the *Nautical Almanac* and brother-in-law of Charles Green. In terms of instruments, instructions, and personnel, this voyage followed the model of earlier ones. In terms of inserting astronomers and expectations of precision onto a naval voyage of exploration that lacked a primary astronomical purpose like a transit of Venus or the initiative of the Royal Society, it helped set the pattern for the future.

The Minutes of the Board for 14 December 1771 list the instruments to be borrowed, purchased, and loaned by the Board for the two ships. It is worth giving this list in full, for it is so close to those prepared by Maskelyne for subsequent voyages that we might see it as defining the instruments of precision exploration at this period. The President and Council of the Royal Society confirmed their willingness to lend:

1 Astronomical Quadrant of One foot radius
1 Astronomical Clock
1 Transit Instrument
2 Common Brass Hadley's Quadrants

Bracketed as being 'Already in the Board's possession' were:

1 Astronomical Quadrant – 1 foot radius

---

38   Confirmed Minutes of the Board of Longitude, 28 November 1771, RGO14/5, 207 (http://cudl.lib.cam.ac.uk/view/MS-RGO-00014-00005/211).
39   Ibid., p. 208 (http://cudl.lib.cam.ac.uk/view/MS-RGO-00014-00005/212).

1 Astronomical Clock
1 Alarum Clock
2 Reflecting Telescopes
Mr Kendal's Watch
Mr Arnold's d^o [Ditto]

The two watches were bracketed together as 'now at the R^l Observatory at Greenwich', the place at which longitude timekeepers were trialled and rates determined. Instruments yet 'To be purchased' were:

2 Journeymen Clocks
1 Alarum Clock
2 of Dollond's last improved 3½ feet Telescopes wth Object Glass
 Micrometers & moveable wires
2 Brass Hadley's Sextants with Mr Maskelyne's Improvements
6 Variation Charts
2 Marine Barometers
4 Common d^o
6 Thermometers
2 Theodolites
2 Wooden frames with Glass roofs for observing by reflection
2 large magnetic needles to use at Land
2 d^o of old construction – to use at sea
2 Magnetic variation compasses
2 Gunter's Chains with spare Links & Rings

Maskelyne was 'desired to bespeak those necessary to be purchased of the proper Instrument Makers and get Ruby Pallets put to the two Astronomical Clocks by Mr Arnold as proposed'.[40] This reveals that Maskelyne was still the chief intermediary with instrument makers, that the Board and the Royal Society both now owned a number of instruments that could be put to repeated use, and that the group of instruments owned by the Board – a government body – was about to be improved and to grow considerably.

At the following meeting there was an addendum to the list, with:

1 Journeyman Clock
2 pair of Globes
A pocket Watch with a second hand } For Mr Wales; Mr Bayly having one
2 of Senex's Maps of the Zodiac
2 large Magnetic Steel Bars for touching the Variation Compasses &
 dipping needles

---

40 Confirmed Minutes of the Board of Longitude, 14 December 1771, ROG14/5, 211–12 (http://cudl.lib.cam.ac.uk/view/MS-RGO-00014-00005/215).

> 12 Copies of the Nautical Almanacs for 1772. 1773. 1774. & as much as may
> be printed of 1775[41]

Such items would not be omitted in future: we might, therefore, consider the printed items and magnetic bars, which serviced other instruments, as much part of the whole 'kit' as clocks and quadrants.

From 1772, very similar lists, with almost identical instructions for their use, were prepared for Board of Longitude astronomers sent on naval voyages of exploration. One was written for Israel Lyons (who had been a computer for the *Nautical Almanac*) in advance of his voyage with Constantine Phipps toward the North Pole, and laid before the Board of Longitude by Maskelyne on 24 April 1773.[42] It shows the same key instruments and books, with the addition of new instruments to be tested. The Board's role in developing and testing new technologies made these voyages part of the process whereby instruments crossed over from being objects under observation to those stable enough to produce trusted navigational information. This happened repeatedly, from the zenith sector at St Helena, to the use of Dollond's achromatic lenses for transits of Venus and the Board of Longitude's trialling of timekeepers, compasses and other instruments and systems. In the case of the Phipps voyage new or improved devices included: 'A Brass Hadley's Sextant, by M\`r\` Dollond, with M\`r\` Maskelyne's & M\`r\` Dollond's new improvements' and Edward Nairne's new marine barometer.[43]

The list that Maskelyne drew up in 1776 for Cook's third voyage (Figure 1.1) is almost identical, as was that for the use of William Dawes, sent as astronomer on *Sirius*, flagship of the First Fleet, to Australia in 1787.[44] These lists of instruments for the use of Board of Longitude astronomers represented only part of the 'instruments of exploration' taken on any vessel. As well as natural history equipment, also missing was equipment typical of Navy ships: octants, ordinary compasses, speed logs, telescopes and more that might belong to individual officers or the ship – and basics like clothing: for the northern voyage in 1773, the Navy

---

41   Confirmed Minutes of the Board of Longitude, 25 January 1772, ROG14/5, 217 (http://cudl.lib.cam.ac.uk/view/MS-RGO-00014-00005/221).

42   Confirmed Minutes of the Board of Longitude, 24 April 1772, RGO 14/5, 238 (http://cudl.lib.cam.ac.uk/view/MS-RGO-00014-00005/242). A fair copy of the list is reproduced and transcribed in Sophie Forgan, Ann Savours, and Glyn Williams, *Northward Ho! A Voyage Towards the North Pole 1773* (Whitby: Captain Cook Memorial Museum, 2010), 56.

43   Forgan, Savours, and Williams, *Northward Ho!*, 56.

44   'List of instruments to be delivered to Capt. Cook May 22 1776', Correspondence of Nevil Maskelyne, AGC/8/29, National Maritime Museum (http://cudl.lib.cam.ac.uk/view/MS-AGC-00008-00029/5); 'Instruments & Books belonging to the Board of Longitude shipped on board the Sirius Man of War bound to Botany Bay' [received 16 February 1787], Papers on the loan of instruments, RGO 14/13, f. 161r (http://cudl.lib.cam.ac.uk/view/MS-RGO-00014-00013/271).

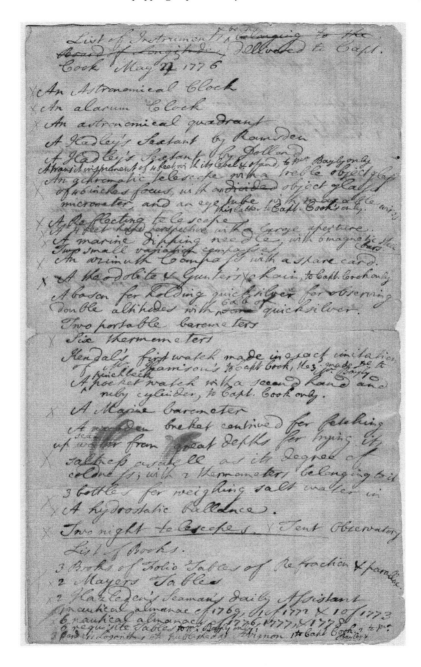

Figure 1.1 Maskelyne's draft list of instruments to be taken on Cook's third voyage

Source: AGC/8/29, National Maritime Museum (http://cudl.lib.cam.ac.uk/view/MS-AGC-00008-00029/5).

Office supplied Phipps's crew with thick clothes, mittens and boots.[45] Instruments supplied by the Navy Board rather than the Board of Longitude, for more ordinary surveying work, also reached ships by other routes. We may assume an influence from one list to the other, as precision instruments became more widely available and, via Maskelyne's expeditionary astronomers as much as anyone else, became more familiar. Officers on Cook's voyages – most famously William Bligh and George Vancouver – would have developed expectations of how voyages should be equipped and observation practiced, so influencing their future requirements.

Maskelyne's selection of instruments for Cook's first voyage sat alongside those which Cook requested from the Admiralty, 'to make surveys of such parts as His Majesty's Bark the Endeavour under my command may touch at'. His request included a 'Theodolite complete', a plane table, a brass scale two feet long, a double concave glass, a 'Glass for traceing the Plans from the light', a parallel ruler, 'A Pair of Proportional Compass's', stationary and Knight's Azimuth Compass.[46] This list is distinctly different in nature but, by the 1790s, it appears that the Admiralty could supply high quality equipment. An example that illustrates this is the list of instruments supplied to Vancouver without the involvement of the Board of Longitude, other than its loan of Larcum Kendall's third timekeeper. It included a sextant by Dollond, and an astronomical quadrant, achromatic telescope and dipping needle by Nairne & Blunt as well as more humble items.[47] Yet 'extreme precision' was still a different matter. Once a Board-sponsored astronomer was added, Maskelyne drew up his list of 'the astronomical & other instruments and books, which seem to me proper to be sent out', and forwarded it to Joseph Banks.[48]

As in the 1770s, this list included instruments belonging to the Board, loans from the Royal Society and some purchased items: this equipment remained expensive and rare enough to be used and reused. The two lists for the Vancouver voyage have some overlap in content, but unique to the precision list are a transit instrument and astronomical regulator. These two items, then, perhaps define precision exploration and survey better than others, although it is the sheer quantity of instruments that could be supplied to and carried by ships – allowing, for example, checks on the errors of particular instruments or individual observers – that characterise the lists more completely. The transit instrument and regulator, however, had to be set up on land and could only be used when there was a reasonably lengthy stop, for they required repeated observations for

---

45   Forgan, Savours, and Williams, *Northward Ho!*, 60.

46   Beaglehole, *Life of Cook*, 136.

47   See appendices to Robin Fisher and Hugh Johnston (eds) *From Maps to Metaphors. The Pacific World of George Vancouver* (Vancouver: University of British Colombia Press, 1993), 291–3.

48   Maskelyne to Banks, 23 Feburary 1791, REG00009/37 (http://cudl.lib.cam.ac.uk/view/MS-REG-00009-00037/458; the 'List of Instruments to be sent for the use of the Astronomer on a Voyage to the North-West coast of America' is at http://cudl.lib.cam.ac.uk/view/MS-REG-00009-00037/750).

Figure 1.2   **The portable tent observatory designed by William Bayly,**
in *The Original Astronomical Observations, Made in*
the Course of a Voyage towards the South Pole, and
Around the World **(London, 1777), Plate II**
Source: B2014, National Maritime Museum.

calibration before they could produce useful data. A portable observatory was
thus required, and Maskelyne added that 'a portable tent Observatory' should be
made for Vancouver's astronomer.[49] William Bayly had designed such tents for the
second Cook voyage, replacing the wooden huts used previously (Figure 1.2). The
details of the Vancouver voyage reveal that tents might now be supplied by the
Navy Board, without involvement of the Astronomer Royal. The tent, clock and
transit instrument are not about exploration that requires mobility over distances
on land: they are an aspect of the particular naval scientific enterprise of making
precise measurements overseas.

The difference between this and another type of exploration and its associated
observation is emphasized by the brief correspondence between Maskelyne and
Mungo Park, the Scots explorer of Africa. Park wrote to the Astronomer Royal
for advice after he was 'engaged to travel into the Interior of Africa' in 1794. He
wished only to 'be enabled to lay down the places with some degree of exactness',

---

49   Maskelyne to Banks, 24 February 1791, REG00009/37 (http://cudl.lib.cam.ac.uk/
view/MS-REG-00009-00037/455).

for which purpose 'I have got a Pocket Sextant which I prefered to a Quadrant for being easier carried or conscealed [sic] and not to liable to accidents'.[50] He had found that 'frequently difficulties arise where we least expect them'; in this case the height of the midday Sun in near equatorial locations was too high to be measured with a sextant. He therefore sought Maskelyne's opinion on his method, which he described briefly and sketched, of observing it indirectly via a reflection (Figure 1.3).[51] Maskelyne's reply discussed the problems of Park's proposal and suggested different approaches. He included considerable detail, giving the impression that he understood the realities of observing in the field with such an instrument. Typically, however, all his suggested methods multiplied Park's equipment and work. Even though 'The observations may be calcu[lated] after your return', he 'should have a good watch w[ith] a second hand', and 'You will doubtless take out the nautical almanac and requisite Tables'.[52] Travelling Maskelyne-style did not come light.

The Expeditionary Astronomers

Maskelyne's instrument lists, expectations and instructions – drawn up after personal experience on scientific expeditions – came to define the work of expert physical observers on Admiralty-sponsored voyages. They also helped make precision observation a defining characteristic of these voyages, first with the expectation of non-naval personnel being included, and, increasingly, being taken on by officers specialising in survey work. For Maskelyne, however, the selection of astronomers for voyages was crucial. They were entrusted with care of the government- or Royal Society-owned instruments, as well as with making the required observations.[53] Maskelyne's astronomers were usually a known and trusted quantity, having been assistants at Greenwich or calculators of the *Nautical Almanac*. He dispensed such work as a form of patronage, supporting those he

---

50   Mungo Park to Maskelyne, 1 October 1794, REG00009/37 (http://cudl.lib.cam.ac. uk/view/MS-REG-00009-00037/676).

51   Park to Maskelyne, 1 October 1794, REG00009/37 (http://cudl.lib.cam.ac.uk/view/ MS-REG-00009-00037/677).

52   Maskelyne's draft reply to Park, dated 4 October 1794, is written on Park's letter (http://cudl.lib.cam.ac.uk/view/MS-REG-00009-00037/677).

53   The observers were to make daily observations for latitude and longitude (by lunar distance and watch), of compass variation and magnetic dip, of air temperature and of the salinity and temperature of the sea. Additional duties might include trialling new instruments, remarking on geographical phenomena and, as often as possible, teaching officers how to use the instruments and new techniques. Establishing positions of headlands, islands and harbours was a basic duty, all of which could be undertaken with greater care if on land for a period of time, where they could also undertake observations of the clock's pendulum to establish the effects of local gravitation: see Howse, 'Principal Scientific Instruments', 120, 124–5, 131.

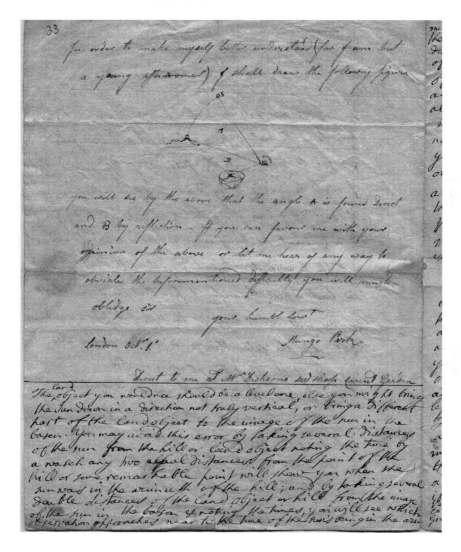

Figure 1.3   Second page of a letter from Mungo Park to Nevil Maskelyne, 1 October 1794, showing Maskelyne's draft reply

Source: Maskelyne Papers, REG00009/37, NMM; http://cudl.lib.cam.ac.uk/view/MS-REG-00009-00037/677..

found capable and for whom he felt some responsibility, often over a long period. For example, once acquainted with Maskelyne, Wales was appointed as one of the first *Nautical Almanac* computers in 1765, as an observer in Hudson's Bay for the 1769 transit of Venus and, in 1772, for Cook's second voyage. Once home, he secured the position of Master of the Royal Mathematical School at Christ's

Hospital. On his last voyage, Wales named the Maskelyne Islands in the south Pacific, recording that he had 'ventured to call them by the name of a person to whom I owe very much indeed; one who took me by the hand when I was friendless, and never forsook me when I had occasion for his help'.[54]

Where Maskelyne did not have personal knowledge, some form of assessment and/or training was required. For example, William Gooch, who was suggested as astronomer for Vancouver's expedition by the Cambridge mathematician Samuel Vince, had to spend several weeks at Greenwich. Gooch reported that although Maskelyne was 'perfectly satisfied with Vinces word', he nevertheless 'attended closely to every thing I undertook for Practice; and observ'd the accuracy of my observations by seeing what they were & calculating what they should be & then seeing how near they agreed'.[55] Gooch was asked to compute the rate of the going of the clock in the Observatory from his own observations, seeing how nearly his results compared with those deduced from Maskelyne's observations. Beyond astronomy, Maskelyne also supplied orders and information. When a cat knocked a timekeeper off a shelf as a voyage prepared to leave, Maskelyne dispatched the maker, Thomas Earnshaw, to Portsmouth with a replacement.[56] Gooch also reported that Wales, who 'pays me very particular attention indeed', gave advice about money, clothes and furniture.[57]

By this date, Banks was the key figure in negotiating the relationship between the Admiralty and the scientific community, but Gooch's letters show how involved Maskelyne remained with regard to physical observation. As he explained to his parents: 'I've requested Vince to give in my name to D[r] Maskaline, so probably before you receive this I shall be nominated astronomer by S[r]. Joseph Banks (President of the Royal Society)'.[58] The nomination came via Banks, but he acted on Maskelyne's advice, despite initially opposing his choice. Similarly, the circumnavigation of Australia by Matthew Flinders is seen as Banks's project, but the selection of observer and instruments remained Maskelyne's business. In 1801 he selected John Crosley, who had been his assistant at Greenwich, the observer

---

54　William Wales, 'Introduction', in William Wales and William Bayly, *The Original Astronomical Observations, Made in the Course of a Voyage towards the South Pole, and Around the World* (London: J. Nourse; J. Mount and T. Page, 1777), lv.

55　William Gooch to his parents, 20 April 1791, and his mother, 29 April 1791, Letters Memoranda and Journal containing the History of Mr Wm. Gooch, CUL MS Mm.6.48, ff. 26–7, 31–2, Cambridge University Library (http://cudl.lib.cam.ac.uk/view/ MS-MM-00006-00048/7). See Richard Dunn, 'Heaving a Little Ballast: Seaborne Astronomy in the Eighteenth Century', in Marcus Granato and Marta C. Lourenço eds. *Scientific Instruments in the History of Science: Studies in Transfer, Use and Preservation* (Rio de Janeiro: Museu de Astronomia e Ciências Afins, 2014), 79–100; Greg Dening, *The Death of William Gooch: A History's Anthropology* (Honolulu: University of Hawaii Press, 1995).

56　Gooch to his parents, 31 July, 7 August and 13 August 1791, CUL MS Mm.6.48, ff. 56v, 61v, 64r.

57　Gooch to his parents, 9 June 1791, CUL MS Mm.6.48, f. 34r.

58　Gooch to his parents, 26 February 1791, CUL MS Mm.6.48, f. 23v.

that replaced Gooch on the Vancouver expedition and, later, a *Nautical Almanac* computer. Crosley in fact left the *Investigator* due to ill health and, in 1802, Maskelyne attempted to appoint the Scottish astronomer Andrew Mackay as a replacement before Mackay's stalling forced him to appoint James Inman instead. These delays meant that ultimately the able Flinders and his brother undertook observations.[59] Flinders had previously travelled with William Bligh, Master on Cook's third voyage, when Bayly was the astronomer. Like Wales, Bayly became a teacher, as head master at the Royal Academy and Royal Naval College at Portsmouth.[60] Such lineages of personnel, instruments and their use can be traced, from the 1760s to the 1800s, to Nevil Maskelyne.

Conclusion

The British tradition of precision exploration outlined here was initiated and coloured by a series of voyages with specific astronomical, navigational and geodetic missions. These were ambitious ventures regarding the production of precise data in overseas locations. The chosen instruments, methodology, and personnel reflected this. Maskelyne's personal experience at St Helena, while not involving exploration of new territories, gave him experience of portable and mobile observatories, and allowed him to pursue the idea of using one opportunity to contribute to several projects. Maskelyne responded to the interests and ambitions of Royal Society and Admiralty circles, and to the model of earlier French expeditions, but much of the technical side – what to observe, how to do it, what with, who by, how to process the data – was left to and developed by him. Maskelyne's lists reveal the development of a standard set of instruments for such voyages, varied perhaps in relation to particular projects, or geographical areas, and a more fluctuating set of instruments as test subjects. They also reveal an overriding respect for certain makers, suggesting the extent to which the maker's name could aid the credibility of the observations.

The lists remind us of different types of instrument for varied uses. There were those for use on board ship, or short stops on land. Where these included novel technologies, still on trial, they are visible, becoming steadily less obvious in the written record as they became more normal parts of a ship's equipment. The lists included instruments that were more expensive and that aimed at levels of precision only possible when established on land and used for a period of time. Individual instruments of this type, such as astronomical regulators and quadrants,

---

59 On Mackay and Maskelyne, see Alexi Baker, '"Humble servants", "Loving Friends", and Nevil Maskelyne's Invention of the Board of Longitude', in Higgitt, *Maskelyne*, 203–28 and 223–4.

60 On Wales and Bayly, see entries in the *ODNB*, and Mary Croarken, 'Tabulating the Heavens: Computing the Nautical Almanac in 18th-Century England', *IEEE Annals of the History of Computing* 23 (2003): 48–61.

once purchased by the Royal Society or Board of Longitude, were used on many occasions. They contrasted with another set of instruments for use on land: the hand-held instruments, chains and rods for measuring on the ground. These were less often the business of the Astronomer Royal, being supplied directly by the Navy Board. The lists also prompt us to recall other items to which we might consider extending the term 'instrument', or which, at least, we must see as essential to the use of precision instruments: magnets to re-magnetise compass or dipping needles, troughs of mercury for artificial horizons, and so on. Certain tables, charts, and navigational manuals were essential enough to be listed; again, we might think of them as instruments, certainly as part of the paraphernalia of expeditionary observation.

This chapter has made a case for the significance of Maskelyne's contribution to eighteenth-century voyages of scientific exploration. This was the result not just of his position as Astronomer Royal but from his previous experience, the institutional links that he embodied, and the moment at which the use of precision instruments was a novelty to be mediated by experts. In the nineteenth century, much of this expertise was internal to the Navy and, following the closure of the Board of Longitude in 1828, it became the business of the Hydrographic Office and of naval academies. Maskelyne's expectation was that the best instruments would be used by trained, mathematically capable observers, asked to take repeated observations whenever opportunity was presented, to achieve the highest possible levels of accuracy. While reality generally failed to meet the ideals implicit in the lists of instruments and instructions for their use, Maskelyne's work helped foster the approach of the British Navy to survey expeditions, provided elements of 'Humboldtian Science' *avant la lettre* and the background to the work of scientific servicemen such as Robert FitzRoy and John Lort Stokes in the decades following.

Chapter 2

Instrumenting Order: Longitude, Seamen and Astronomers, 1770–1805

Eoin Phillips

On the morning of 28 April 1789 in the South Pacific, William Bligh of HMS *Bounty* was woken by his Master at Arms and recently-promoted Acting Lieutenant, Fletcher Christian. These two ranking members of the ship's crew proceeded to bind Bligh's hands with cord and threaten him with death should he make a sound. With the aid of two more crew, Bligh was carried on deck: he found 'no man to rescue me'. In his reconstruction of the event in his *A Voyage to the South Sea*, Bligh remembered the scene:

> The boatswain and seamen who were to go in the boat were allowed to collect twine, canvass, lines, sails, cordage, an eight-and-twenty gallon cask of water; and Mr. Samuel got one hundred and fifty pounds of bread, with a small quantity of rum and wine, also a quadrant and compass; was forbidden, on pain of death, to touch either map, ephemeris, book of astronomical observations, sextant, time-keeper or any of my surveys or drawings … (Mr. Samuel) attempted to save the time-keeper, and a box with my surveys, drawings, and remarks for fifteen years past, which were numerous; when he was hurried away, with "Damn your eyes you are well off to get what you have".[1]

The time-keeper in question, known now as 'K2', had been supplied for the voyage by the Board of Longitude. K2 had been bought by the Board from the watchmaker Larcum Kendall in 1772, its second commission to Kendall. The task was to make a less expensive version of Kendall's 'K1' – Kendall's replica of John Harrison's much-celebrated timekeeper now known as 'H4' – although K2 was still expensive, costing the Board £200.[2]

After taking the timekeeper and forcing Bligh and some of his followers into a cutter, the mutineers sailed to Tahiti, which the *Bounty* had left 23 days previously. Tahiti was a crucial site for the voyage of the *Bounty*, not least because Bligh had been tasked with collecting breadfruit plants from there. The *Bounty* was specially fitted to carry back large quantities of breadfruit to England, part of a

---

1 William Bligh, *A Voyage to the South Sea* (London: George Nicol, 1792), 156–7.

2 Derek Howse, 'Captain Cook's Marine Timepieces Part 1: The Kendall Watches', *Antiquarian Horology* 6 (1969): 190–92.

Royal Society plan, headed by Joseph Banks, to replant the breadfruit in the West Indies in order to provide cheap food for slaves and to lessen the need to rely on trade from the United States of America. A party of the mutineers, along with a number of Tahitian men and women, promptly left Tahiti to settle on Pitcairn Island because, they believed, Pitcairn had been wrongly charted on Royal Navy maps. This fact, together with the burning of the *Bounty* would, they hoped, make them untraceable. Bligh and most of his exiled crew did manage to return to England, in what has become a much-vaunted 47 day voyage to the Dutch-held island of Timor.

In 1974, W.E. May celebrated the return of K2 to the Royal Observatory at Greenwich that had secured it in 1969.[3] The importance placed on 'the return' of the timekeeper to Greenwich is indicative of prevailing understandings of early British activity in the Pacific, wherein great importance has been placed on technologies such as timekeepers as both the tools for enabling such voyages and the artefacts that define their history. In 2010, a marine timekeeper made by Thomas Earnshaw was included in Neil MacGregor's *A History of the World in 100 Objects*. The justification for the inclusion of Earnshaw's No. 509 was not only that it typified the advanced craft skill that typified English technology in the Enlightenment, but also that it was taken on several voyages of exploration, including Fitzroy's 1831 HMS *Beagle* voyage. As such, it represented the impressive technology that led to the kind of accurate navigation and precise mapping that allowed Britain to have such political and economic reach over much of the globe.[4]

One result of the attention paid to European-made timekeepers and astronomical instrumentation was that the long-range practices of empire and of precision science have been reckoned inherently secure. Metropolitan centres like London produced the hardware to effect this. This view corresponds, in turn, to a general characterisation of science and empire in which both have been seen to be 'made' in the centre and then by various means spread 'outwards' across the world.[5] This interpretation has had particular resonance for histories and geographies of the Enlightenment, in which the instruments, charts and maps of European enterprises have been characterised as embodying and creating the virtues of rationality and order.

William Bligh knew well the delicacy of the practices on which his enlightened work depended. While on Tahiti in January 1788, Bligh noted that the expensive K2 with which he had been entrusted was vulnerable to local circumstances:

---

3   W.E. May, 'How the Chronometer Went to Sea', *Antiquarian Horology* 9 (1974): 640.

4   Neil MacGregor, *A History of the World in 100 Objects* (London: Allen Lane, 2010).

5   George Basalla, 'The Spread of Western Science', *Science* 156 (5 May 1967): 611–22. For a critique of Basalla and as an example of an alternative to this kind of model, see Kapil Raj, *Relocating Modern Science: Circulation and the Construction of Knowledge in South Asia and Europe, 1650–1900* (New York: Routledge, 2007).

In consequence of my having been kept all night from the ship by the tempestuous weather, the time-keeper went down at 10h 5" 36'. It's rate previous to this, was 1",7 losing in 24 hours, and it's error from the mean time at Greenwich was 7' 29"2 too slow. I set it going again by a common watch, corrected by observations, and endeavoured to make the error the same as if it had not stopped; but being over cautious, made me tedious in setting it in motion, and increased the error from mean time at Greenwich.[6]

More important for Bligh, however, was securing the safety of his journal. In his account of the voyage after the mutiny, Bligh emphasised the role of John Samuel in securing his papers: 'Without these, I had nothing to certify what I had due, and my honour and character would have been in the power of calumny without a proper document to have defended it'.[7]

The need for Bligh to rely on written documents in order to secure his honour points to the difficult relationship that existed in the late eighteenth century between trust, visibility, and discipline. The mutineers, too, knew well the connections between the discipline and what it meant to be visible. The existence of bodily violence was no less real for sailors on European voyages to the Pacific. As well the constant threat of the lash, sailors understood that the odds on returning to Europe were low. Greg Dening has suggested that more than 40 per cent of European sailors who entered the Pacific in the eighteenth century never returned.[8] The incidence of mutiny in the late eighteenth century has been understudied, but recent accounts suggest that 'at least one third of European warships experienced some form of collective rebellion during the 1790s'.[9] The grid that is made up by the lines of longitude and latitude and the instruments related to them – such as timekeepers – have come to stand as symbols of rationality and enlightenment. Fletcher Christian and his fellow mutineers hoped to go 'off grid', to make themselves invisible. These facts of visibility and of discipline are important. So, too, was the fact that the instruments and paper on which such facts and lines relied were delicate, prone to breaking and to damage. This was of central concern to the British Admiralty and to the Board of Longitude in their attempts to develop a solution for finding longitude at sea, particularly when the types of instruments being tried on these voyages were both expensive and difficult to use.

The tension that existed between the ambition of the Board and the aims of metropolitan science and technology against the routine practices of shipboard discipline has too often been conflated within accounts which tell of the singular

---

6 Bligh, *A Voyage to the South Sea*, 119.

7 Greg Dening, *Mr Bligh's Bad Language: Passion, Power and Theatre on the Bounty* (Cambridge: Cambridge University Press, 1992), 106.

8 Ibid., 141.

9 Clare Anderson, Niklas Frykman, Lex Heerma van Voss and Marcus Rediker (eds) *Mutiny and Maritime Radicalism in the Age of Revolution: A Global Survey* (Cambridge: Cambridge University Press, 2014), 4.

success of European missions to the Pacific. In accounts of how timekeepers went to sea, a group of personnel has similarly been made invisible: I refer here to that group of astronomers sent by Greenwich to use and test timekeepers alongside other expensive instruments. In examining the work of these astronomers on voyages of exploration, this chapter aims to illuminate the tensions that existed between navigation, instrumentation, discipline, and reporting. The chapter makes two claims. First, in putting clocks and watches to sea, problems of discipline, disciplinarity, and orientation were raised but not always solved. The second is that the solution for discipline and visibility rested more on successful performance than it did on the functioning of the hardware itself.

Astronomers at Sea

Between 1772 and 1803 the Board of Longitude commissioned several men to travel as astronomers on state-sponsored voyages of exploration. Each astronomer was equipped with a large and expensive package of instruments and texts and each carried with him a set of standardised instructions written by the Astronomer Royal, Nevil Maskelyne[10] (Table 2.1).

Table 2.1   Nevil Maskelyne's instructions to astronomers on voyages of exploration[11]

Observations to be made on shipboard

1. You are everyday, if the weather will permit, to observe meridian altitude of the sun, for finding the latitude and the altitudes of the sun, both in the morning and afternoon, at a distance from noon, with the time between measure by a watch, and the sun's bearing by the azimuth compass at the first observation, in order to determine both the apparent time of the day and latitude, for fear the sun should be clouded at noon: you are moreover to observe distances of the moon from the sun & first stars, with the Hadley's sextant, from which you are to compute the longitude of the ship by the nautical almanac.
2. You are to wind up the watches every day, as soon after the time of noon as you can conveniently, and compare them together, and set down the respective times; and you are to note also the times of the watches, when the sun's morning and afternoon altitudes or the distances of the moon from the sun & fixt stars are observed; and to compute the longitude resulting from the comparison of the watches with the apparent time of the day inferred from the morning and afternoon altitudes of the sun.
3. You are to observe or assist at the observation of the variation of the compass, and to observe the inclination of the dipping needle from time to time.

---

10   For a list of instruments supplied by the Board of longitude, see Cambridge University Library Royal Greenwich Observatory [hereafter CUL RGO], MS 14/68, 173.

11   CUL RGO 4/185, 'Instructions for John Crosley to go on a voyage to New Holland on board HMS 'Investigator' on a scientific expedition, 7 March 1801', 4v-4a(r).

4. You are to note the height of one or more thermometers places in the air, and in the shade, early in the morning, at about the hottest time of the day; and to observe also the height of the thermometer with the vessels, near the watches; and to make experiments of the saltness of the sea; and the degree of cold by letting down the thermometer to great depths; as you shall have opportunity; and to make remarks on the southern lights, when you are far to the south; if any should appear.

5. You are to keep a ship-journal, with the log worked according to the dead reckoning, allowing for lee-way and variation; noting therein the length of the log-line; and the time of running out of the sand-glasses from time to time, by the help of the watches; and you are to insert therein also another account corrected by the last celestial observations, and a third deduced from the watches.

6. You are to teach such officers on board the vessel as may desire it, the use of Hadley's sextant in taking the moon's distance from the sun and fixt stars, and the method of computing the longitude from the observations.

7. You are to settle the positions of head-land, Islands, and harbors, in latitude and longitude, by the celestial observations; and also set down what longitude the watches give to them.

Observations to be made on shore

1. Whenever you land, if time permits, you are to set up the tent Observatory, and astronomical clock; either setting the pendulum exactly to the same length as it was of at Greenwich, before the voyage or noting the difference by the revolutions of the screw and divisions of the nut at bottom; & you are to take equal altitudes of the sun & fixt stars with the universal theodolite, or with Hadley's Sextant by reflexion from the surface of quicksilver in a bason according to circumstances, for determining the rate of going of the clock, and for finding the time of noon, and fixing a meridian for the universal theodolite, which is after that to be used as a transit instrument & meridian circle. You will thus, also, by the difference of the going of the clock at the place, from it going at Greenwich before the voyage, determine the relative force of gravity at the two places.

2. You are to wind up the watches, everyday, soon after the time of noon as you can conveniently, and compare them together, and with the astronomical clock, at that time, and also about the times of equal altitudes, if any were taken.

3. You are to observe the meridian altitudes of the sun, & also of the brighter stars some to the north & some to the south, for finding the latitude with the universal theodolite, or with the Hadley's sextant by reflection from the bason of quicksilver, according to circumstances; & note the height of the thermometer, and that of the portable barometer.

4. You are to observe continually, transit of the sun & brighter fixt stars, begin of nearest to the equator, over the meridian … with the universal theodolite, and to observe eclipses of Jupiter's first satellite, and occultations of fixt stars by the moon; & to take distances of the moon from the sun ad fixt stars, with Hadley's Sextant from all which you are to compute the going of the clock and watched the longitude of the place, by the nautical almanac.

5. You are to observe the height of the tides, and the times of high and low water, particularly at the full and change, and quarters of the moon; and to note whether there be any difference, and what, between the night and day tides.

6. You are to observe the variation of the compass, and the inclination of the dipping needle.

7. You are to note the height of one or more thermometers, places in the shade, early in the morning, and about the hottest time of the Day.

Among these demanding and seemingly individuated tasks, onboard instruction No. 6 stands out as peculiarly interactive. This instruction 'to teach' represented the ambition of the Astronomer Royal to promote the use of the 'lunar-distance method' for finding longitude. In particular, Maskelyne was keen to promote the use of the *Nautical Almanac*, which was first published in 1767, its contents based on the intensive observation records compiled by astronomers and astronomical assistants managed by the Astronomer Royal at Greenwich. This described the position of a range of celestial bodies for each hour of a year and their position relative to the earth's surface.

A problem for astronomers on voyages, and for the Board of Longitude, was getting a ship's crew, particularly its officers and midshipman, near to an adequate operational standard such that they could be trusted to perform the required instrumental observations. At the end of the logbook compiled whilst serving as astronomer on Cook's second voyage, the astronomer William Wales complained that even the idea of the civil and the astronomical day caused immense confusion amongst the crew – astronomical days started and ended at midday, rather than at midnight: 'It is much to be lamented', exclaimed Wales, 'that Seamen and Astronomers should not reckon their days alike, as this difference is very troublesome, and causes much confusion, especially amongst younger hands, in looking for and taking things out of the Nautical Almanac. I have met with some whom I never could make fully comprehend it'.[12]

For Wales, the importance of teaching lunars was clear. As long as seamen and officers remained confused by basic astronomical and mathematical practices and language, then neither timekeepers nor Nevil Maskelyne's *Nautical Almanac* project could be deemed valid in solving the longitude problem. The sense that the Navy was uneducated in the terms of astronomical routine, and that this was a problem, was felt not just by the Astronomer Royal and the Board of Longitude. John Brisbane, astronomer on an East India Company ship bound for Parramatta, recalled in 1795 how 'in that immense fleet there were perhaps not 10 individuals who could make a lunar observation'.[13]

A Naval Academy had existed in Portsmouth from 1733. From 1807, it existed as the Royal Naval College, and was headed by James Inman, formerly astronomer on Matthew Flinders' 1801 HMS *Investigator* voyage. The general aim of the Academy and of the College was to train officers to become proficient in taking lunars and computing the results, so that they could be trusted to use the *Nautical Almanac* and so check the rates of clocks reliably. Alongside this astronomical education, naval students also received a general education in disciplines such as mathematics and drawing. There was, however, a very small intake of students.

---

12   CUL RGO 14/58, 'William Wales Log Book of HMS Resolution, 1772–5', 187.

13   Quoted in Simon Schaffer, 'In Transt: European Cosmologies in the Pacific', in Kate Fullagar (ed.) *The Atlantic World in the Antipodes: Effects and Transformations since the Eighteenth Century* (Newcastle upon Tyne: Cambridge Scholars Press, 2012), 70–93, quote from page 84.

For the lifespan of the Naval Academy, 98 per cent of the Royal Navy's officer entry proceeded to sea via the long-held route of family and social connections. For the period in which the Royal Naval College at Portsmouth operated, only about 10 per cent of officers went to sea as graduates: the great majority proceeded according to traditional routes of entry. For Dickinson, the reason for this rejection of the naval schools and academies was clear, and clearly expressed at the time: 'officers saw it as an attempt to wrestle the powerful and very ancient vested interest – the right of senior officers to choose their entourage and those who would ultimately succeed them'.[14] Thus, the traditional system of master and apprentice was regularly reproduced on voyages of exploration. William Bligh, for example, served as sailing master on Cook's third voyage, and was joined by George Vancouver; Matthew Flinders had been a midshipman on Bligh's second breadfruit voyage. As a representative of the Board of Longitude and the Royal Observatory at Greenwich, Maskelyne was keen to have an astronomer present on voyages of exploration, but this position needed to sit well with established hierarchies of command on board ship.

Astronomers were not guaranteed to have authority on board, yet their ability to teach officers relied upon their having an authoritative position. In March 1773, during Cook's second voyage of exploration, the *Resolution* was headed towards Dusky Bay. On 11 March, William Wales recalled in his logbook how Cook and his fellow officers had challenged his hypothesis to account for the dramatic variation in compass readings: 'Having frequently remarked on very considerable irregularities in the Observed Variations of the Compass, I sometime ago took occasion to examine into the circumstances under which they were made. ... I mentioned this to some of the Officers and also to Capt. Cook, who said I was a Philosopher [original emphasis]. However this morning, the following [readings] were taken (I suppose) to confute me'.[15] The label 'philosopher' was here used by Cook to denote someone who made bold claims based on spurious evidence: Cook, the established seaman, was making a clear difference between the practical skill of the experienced naval officer and the land-based training of the Greenwich-approved astronomer.

Securing the trust of officers mattered to astronomers not simply because officers had to be taught, but in order that the instructions given to the astronomers could be attempted. Francis Reid has argued that a large part of Wales' success as an astronomer resulted from his ability to forge good working relations with the Royal Navy officers with whom he travelled.[16] For Reid, the relatively large cabin given to Wales on the voyage was an indication of his status and a way to maintain authority on the ship. On a Royal Navy vessel, social hierarchy was

---

14   Quoted in H.W. Dickinson, *Educating the Royal Navy: Eighteenth- and Nineteenth-Century Education for Officers* (London: Routledge, 2007), 45.

15   CUL RGO 14/58, 'William Wales Log Book of HMS Resolution 1772–5', 33.

16   Francis Reid, 'William Wales (*c.*1734–1798): Playing the Astronomer', *Studies in History and Philosophy of Science* 39 (2009): 170–75, quote from page 171.

strictly enforced and embedded in the architecture of a ship: the size of a cabin reflected the status of its occupant.

The need to achieve success through others' careful management of social status onboard was an issue which affected all the longitude astronomers. Before agreeing to depart on his second voyage of exploration, astronomer John Crosley told Maskelyne that 'he [had] no objection, if his situation can be made comfortable'. Crosley was alluding, Maskelyne explained to Banks, 'to his situation on board Capt. Broughton's Ship, where he had room enough at first, but the Captain having afterwards contracted the room, his hammock was continually knocking against the partition, which disturbed his sleep & [was] injurous to his health'.[17] Even when astronomers were content with their quarters, it was not at all clear, at least from their instructions, how they could always access instruments such as their timekeepers. On Cook's second voyage, the timekeepers were not kept with their operating astronomer, but were instead stored in the Great Cabin.[18] On Crosley's first voyage on HMS *Providence,* he wrote to Maskelyne alluding to the complications that had arisen from having his timekeepers 'fixed' in the Great Cabin: 'I carried my Sextant, artificial Horizon and a Pocket Watch No 1522 (made by Mr Earnshaw for the purpose of being carried with the Time keepers to take the time on Deck or on shore without moving the Time-keepers which are fixed in the Great Cabin, which does not alter its rate above 5 or 10' per day).[19] As Crosley hinted at in this letter, moving instruments was something he was made to feel – by some clock and watchmakers anyway – that was to be avoided. Yet, as was clear from the 'on shore' instructions given to astronomers (Table 2.1), moving timekeepers was a necessity. That some timekeepers had to be moved was a fact built into their mechanism. The watches made by the London watchmaker Thomas Earnshaw even required a particular movement of the wrist to set the watch going – as Earnshaw instructed his customers. In a letter to Earnshaw in 1805, Captain Archibald Swinton was careful to report that, after his timekeeper had stopped, he had proceeded to handle the timekeeper correctly, setting it going by 'giving it a circular motion'.[20] As noted above, William Bligh thought it was by 'being over cautious [that] made me tedious in setting it in motion, and increased the error from mean time at Greenwich'.[21]

---

17   CUL RGO 35/55, 'Letter from Maskelyne to Banks recommending Crosley to serve as astronomer onboard the Investigator, 23 January 1801'.

18   Reid, 'William Wales', 171.

19   CUL RGO 14/68, 'Letter from Crosley to Maskelyne from HMS *Providence*, Plymouth Sound, Dec 20, 1794', 95.

20   'Letter from Captain Swinton to Thomas Earnshaw, December 30, 1805', in Thomas Earnshaw, *Longitude, An Appeal to the Public: Stating Mr. Thomas Earnshaw's Claim to the Original Invention of the Improvements in His Timekeepers* (London: F. Wingrave, 1808), 252.

21   Bligh, *A Voyage to the South Sea*, 119.

For those instruments that were less 'mobile', astronomers were keen to report that when bumps or breakages occurred in using the instruments in their possession, this was despite the care taken to prevent this. In 1791, while preparing for his voyage to join Vancouver's expedition surveying the north-west coast of America, the astronomer William Gooch reported that somehow the Earnshaw timekeeper he was carrying managed to be knocked off the shelf in his cabin on which he had placed it, apparently by the ship's cat. Thomas Earnshaw was 'not amused', having to go post-haste to Deptford from London with another clock.[22] On his maiden voyage as a longitude astronomer, Crosley was especially unlucky. On 16 May 1797, sailing from Macao, the *Providence* struck a coral reef. Crosley reported that though 'every effort was made by the Captain, Officers and Ships Company to get the Ship off the reef ... it was without success'. He then made it clear that he had made every effort to collect his instruments, books and computations made on the voyage. Fortunately, he recounted, he was able to rescue from the gun deck 'the Time keeper No. 1 belonging to the Navy board, my Time Keeper, Hadley's Sextant, Journal, Book in which the computations were made during my voyage and the memorandum books in which the observations were [written] as they were observed on Deck'. But Crosley lost or damaged all these objects when he fell from a ladder leading to the top deck after a sudden wave hit the ship. This accident caused 'the Time-keepers and memorandum books [to] fall out of my Pockets, the Time-Keeper No.1 and the books fell into the launch, my own Time-keeper fell overboard and was lost'.[23]

Forms of rhetorical repair were important strategies deployed by astronomers to explain deviations from their instructed tasks.[24] At the same time, other kinds of repair were routine on voyages. When James Inman eventually reached Flinders and took control of the instruments left by John Crosley, he spent considerable sums on their repair and making provisions lest they should be damaged again. At Port Jackson, Inman paid £3-5-0 for 'repairing Astronomical Instruments, cleaning Sidereal, & assistant Clocks'.[25] At Madeira, Inman paid a further £1-1-0 simply to take Kendal's timekeeper ashore.[26] This was in addition to buying several extra cases for the sextant and the timekeepers, paying £3-10-0 in order to stow his instruments for safe travel in a ship from New South Wales to China, and paying a surcharge for safe transit of the astronomical instruments from Long Reach to London and then to Greenwich. Upon his arrival to Port Jackson to meet Flinders and his crew, Inman paid 10 shillings to a carpenter to have the tent fixed.[27]

---

22   CUL MS. Mm.6.48, 'William Gooch, letter to his parents, 13 August 1791', 64.

23   Ibid., 164.

24   Simon Schaffer, 'Easily Cracked: Scientific Instruments in States of Disrepair', *Isis* 102 (2011): 706–17.

25   CUL RGO 14/1, 'James Inman's Account of Expenses during a Voyage from England to New South Wales and back by the Way of China in 1802, 1803, & 1804', 191.

26   Ibid., 191.

27   Ibid.

Understanding that Board of Longitude instruments got damaged and needed repair between voyages became routine for the Board. This was demonstrated by the frequent, and quite substantial, payments made by the Board to makers like Arnold and Earnshaw. Between the voyages in which their timekeepers had been used, these clock makers were regularly paid to clean and to service the instruments. When equipping the astronomer John Crosley for his voyage with Captain Flinders, John Arnold was paid £6-6-6 for 'cleaning Kendals [sic] third made timekeeper, with two new boxes & gimbols'.[28] Upon the return of the timekeepers from Gooch's ill-fated voyage, Earnshaw was paid five guineas by the Board of Longitude for the cleaning and repair of the timekeepers.[29] The fact that instruments such as timekeepers were subject to damage and that their repair was not considered a major issue is proof that, to the Board of Longitude, timekeepers were simply not expected to work unproblematically once out of the hands of their makers. The instructions to astronomers hint at this in noting that astronomers, whenever they landed, were to check the rates of their timekeepers against complex astronomical procedures which involved the erection of a portable observatory and astronomical clock and undertaking skilled and arduous series of observations.

Yet the shore-based instructions issued by the Board posed several problems which the astronomers routinely reported upon. The time taken to determine the longitude by lunar observations on land against which an astronomer could check the rates of the timekeepers could be considerable. On Vancouver's 1793 voyage, 2,046 single observations were required.[30] The fact that the taking of observations was so arduous and time consuming was of concern not only for the longitude astronomers, but also for the entire crew. Reacting quickly to suitable winds, navigating to avoid poor weather, and ensuring suitable provisions were all of concern to a ship's crew. Time spent on shore constructing an observatory required considerable work. The logbooks of the longitude astronomers are peppered throughout with accounts of the difficulties of finding flat ground, or keeping the instruments dry. In March 1773, for example, William Wales spent over a week felling trees to clear ground for his observatory, only for his observations to be then hampered by cloudy weather.[31]

Other kinds of problem also presented themselves, requiring astronomers to draw on resources well beyond their scripted instructions. Setting up an observatory on the American Atlantic and Pacific coasts often meant negotiating with other Europeans, treading tentatively between diplomacy, hostility, and incommensurable ideas of property. Both William Gooch and John Crosley

---

28   CUL RGO 14/7, 'Minutes of the Commissioners of the Board of Longitude, December 2 1802', 15.

29   Ibid., 16.

30   Alun Davies, 'Horology and Navigation: The Chronometers on Vancouver's Expedition, 1791–5', *Antiquarian Horology* 21 (1993): 247.

31   CUL RGO 14/58.

understood the importance of trying to act as an astronomer in the Spanish-dominated West coast of America. In their logbooks, both men recount the difficulty in negotiating with the Spanish governors in order to establish their observatories. To prepare for his voyage, Gooch spent time with a language tutor. In a letter to his parents, Gooch took excited pride in telling his parents that alongside lessons from Señor Tsola, an Italian with excellent Spanish, he was about 'to begin Don Quiote in the Original. While on Ship Board I shall want some study for Amusement and that I may have a variety, I'll take Latin, French, Spanish & Italian Books, that I may be improving myself in the Classical way or getting a Knowledge of the most useful modern Languages according as I find myself Inclind'.[32]

'Negotiating' and exchange was not always so cordial. Crew members and astronomers hoping to have access to a beach might encounter hostile parties. It was in an astronomer's interest, therefore, to be part of a crew armed with muskets. Relations with indigenous cultures in the South Pacific often required a less bloody interaction. Much has been written about the commercial opportunities of exploiting the passion of Pacific islanders for trinkets. As Nicholas Thomas has stressed, this was not a simple 'gift economy', but one, rather, which set European subservitude in balance.[33] Astronomers were accordingly encouraged to be equipped with a range of iron goods. The value of trinkets and the value of scientific instruments could become conflated, however; and astronomers were also concerned about islanders stealing their instruments. So grateful was he to the diligence of one assistant that James Inman paid £8 as 'a Present to Convict Servant on leaving Port Jackson, for his great Care, & Attention in Watching at the Tent Observatory, when Indisposition prevented me from sleeping at it myself'.[34]

That longitude astronomers persistently encountered problems in these ways begs the question why the Board continued to send them and their range of expensive instruments on voyages of discovery. One answer is that members of the crew simply did not have the astronomical or navigational skills necessary to rate timekeepers or to take the observations required to produce the logs and charts demanded by the Admiralty and the Board of Longitude. Yet several incidents related to the Board's expeditions and its astronomers reveal that officers often did have the skills required to perform the tasks that the Board paid its astronomers to perform. The next part of this chapter will relay one such incident involving the claims of Captain Matthew Flinders and his brother Lieutenant Samuel Ward Flinders, before considering what it was that astronomers seemed to do differently, and what this tells us about the purpose of sending astronomers on voyages of exploration and on the relations between the ambitions of naval officers and those of metropolitan astronomy.

---

32   CUL MS. Mm.6.48, 20.

33   Nicholas Thomas, *Entangled Objects: Exchange, Material Culture, and Colonialism in the Pacific* (Cambridge: Harvard University Press, 2001).

34   CUL RGO 14/1, 'Correspondence relating to payments to James Inman, including £100 for his passage from China'.

Seamanship and Authorship

In October 1801, John Crosley decided to leave his voyage with Matthew Flinders on HMS *Investigator* at the Cape of Good Hope. On 5 March 1802, at a meeting of the Board of Longitude, Crosley's letter was received, stating his reason for leaving the voyage and requesting that a new astronomer be assigned to replace him.[35] By June 1803, the new astronomer, James Inman, had joined Matthew Flinders on the *Investigator*. When Flinders arrived at Port Jackson, in May 1802, he informed Maskelyne that, in lieu of an official astronomer, the tasks of the astronomer were still being undertaken:

> on Mr Crosleys quitting the Investigator, I took upon myself to fulfil the duty of astronomer as far as my knowledge would permit, and as the time which could be spared from surveying and my other avocations would allow. I much fear that you will have good occasion to find fault with the manner in which this duty has been done, both from the little time that I could spare since coming upon the coast of New Holland, as well as from a want of knowledge. I trust that Doctor Maskelyne will excuse a stranger addressing him when the subject is to inform him of what has been done on board the Investigator in the department of astronomy.[36]

In his initial letter to the Astronomer Royal, Flinders commented on the performance of the Earnshaw timekeepers that had been taken on the voyage [Thomas Earnshaw's 542 and 520], but he concentrated mainly on the problems he had encountered in carrying out the astronomical duties that had accompanied Crosley. Two of the timekeepers – Arnold's No. 12 and 176 – had altogether stopped working.[37] In using the astronomical clock, Flinders reported that there was imply had too little time 'to do any thing with it'. Only when in King George's Sound had he been able to set up the astronomical clock in association with Earnshaw's No. 453 (which had been going so well that he decided to make it the standard timekeeper with which to compare the rates of the other timekeepers). He had the intention of making this astronomical clock the standard for all observations in the port, but, unfortunately, the weather had been so bad that no observations could be obtained.[38] It was more than the weather that caused Flinders concern. As he stressed to Maskelyne, 'a great obstruction to our operation' was the smallness of the portable observatory, by which he meant that the theodolite and clock had to stand in different tents. In addition, Flinders complained that the canvas of it

---

35   CUL RGO 14/68, 'Letter from Matthew Flinders to Nevil Maskelyne, from HMS Investigator at Port Jackson, May 25 1802', 16.

36   Ibid.

37   CUL RGO 14/68, 'Letter from Matthew Flinders to Nevil Maskelyne, from HMS Investigator at Port Jackson, May 25 1802', 17.

38   Ibid., 17.

[the tent] 'was rotten and full of holes', arising 'from the little room in the ship, which obliged us to take the parts out of the cases and stow them separately in different places'.[39]

The concerns raised by Flinders are reminders of the problems inherent in making fragile hardware travel and perform satisfactorily on long-distance voyages. What is often less apparent are the problems that existed in managing and deploying the whole host of other hardware that was seemingly less fragile, but upon which the performance of astronomical and navigational duties also often depended. These more veiled problems are revealed to us from an account entitled 'Abstracts of Astronomical Observations', by Matthew Flinders' brother, Lieutenant Samuel Ward Flinders, who had also been on the voyage, and received by the Board of Longitude in September 1804. In 'Abstracts', the junior Flinders presented his observations for finding the longitude of many places around New Holland, as well all his lunars undertaken in what he called his 'Astronomical journey'.[40] The observations, noted Flinders, were all made using a Troughton Sextant No. 4. Flinders deemed it necessary to note the problems inherent in handling this sextant. He stressed that the observations would only very nearly agree provided that great 'pains are taken in the Observ g [sic]'. These 'pains' could only be remedied by 'the Observer himself having no weight to sustain, that the arch of the sext t [sic] is not moved up to the eye to read off; & that the Obs ns are taken within 24 h of each other'.[41]

Although they are not presented in the recognisable arrangement of observations as taken by other astronomers on voyages, Lieutenant Flinders' findings consisted of a multitude of data, with extensive commentary about his observations. Samuel Flinders sent these astronomical results to the Board of Longitude in 1804 and, by virtue of their range and detail, demanded payment at the rate the Board paid its appointed astronomers. The Board was reluctant to so pay him. Between 1805 and 1808, Lieutenant Flinders continued to feel aggrieved at what he regarded as the Board's protracted negotiations over his observations and its refusal to pay him the requested astronomical salary. Unlike Crosley, Flinders had provided the Board with an extensive set of nautical and astronomical observations. Unlike the Board's treatment of Crosley and other astronomers, the Board for its part demanded that Lieutenant Flinders prove four things to them: that he alone filled (completed) the astronomical work; that when it was found necessary to get the tents on shore, that he alone had cared for the instruments, made the observations, and performed the labour of calculation; that he kept the journals of these observations; and, that it was upon Lieutenant Flinders that they could place trust and dependence for what was performed in the 'Astronomical line'.[42]

---

39   Ibid., 18–19.

40   CUL RGO 14/68, 'Abstracts of Astronomical Observations made in the years 1802 & 3 on the coasts of New Holland'.

41   Ibid., 23.

42   Ibid.

In addition to testimony from his brother, Lieutenant Flinders assembled his case by writing to several officers and gentlemen who had been on board. In early May 1806, the Board was presented with statements from senior members of the voyage, including the botanist Robert Brown, the officer Robert Fowler, and the botanical artist Ferdinand Bauer, all of whom answered in the affirmative regarding the four questions put to Flinders by the Board. By April 1808, Lieutenant Flinders still had not received any money from the Board. In a further petition presented to the Board, Flinders stated again that it was he who filled the role – 'department' – of Astronomer on the *Investigator*. It was, he stressed, during this period 'in which all the Discoveries were made, and all the Unknown-coasts explored in that Voyage'.[43] He also emphasised that this duty was 'discharged . . . without assistance; accepting, occasionally a few Latitudes and Variations, and nor wanted then a few Lunar Distances from Captain Flinders, whenever his attention from the Geographical Department would admit'.[44] Though the Board eventually accepted Samuel Flinders' observations, he was granted only £200 and this was paid into his agent's hands, a sum, Flinders recounted, which 'did not amount to the salary of the Astronomer for half a year'.[45] Lieutenant Flinders, simply, was not deemed to have acted as an astronomer.

To understand the actions of Samuel Ward Flinders and the response of the Board of Longitude, it is first necessary to appreciate the circumstances of his brother, captain of HMS *Investigator*. Matthew Flinders understood the social status accorded successful voyaging even before he departed. In a letter in 1801 to Willingham Franklin (brother to the Arctic explorer John Franklin), he boasted that 'should this voyage prove successful, I shall not be unknown in the world; my acquaintance in Soho Square will introduce me to many of the first philosophers and literati in the kingdom'.[46] A successful voyage meant more than just establishing himself as a fixture in the socio-scientific gatherings around Joseph Banks – his 'acquaintance in Soho Square'. It would also mean that he could guarantee his extended family financial support, and perhaps secure for himself full promotion to Captain. Flinders was aware, however, of the need to command 'the perfect voyage'. Greg Denning has noted that this concern for creating the 'perfect voyage' was absolutely central as a motive to the aspiring Lieutenant William Bligh. As well as making him wealthy, commanding 'the perfect voyage' would offer Bligh the chance to gain promotion and propel him in status towards that of Captain Cook, whose apotheosis had rapidly taken place in Europe as well as in Hawai'i.[47]

---

43   Ibid., 33. The *Investigator* ceased surveying on June 9 1803, due to extensive leaking in the ship thought to be caused by rotting wood.

44   Ibid., 34.

45   Ibid., 'Letter from S.W. Flinders to George Gilpin, May 21st 1808'.

46   Ernest Scott, *The Life of Captain Matthew Flinders* (Sydney: Maclehose, 1914), 281.

47   Denning, *Mr Bligh's Bad Language*. On the myths of Cook's fate both in England and in Hawai'i, see Marshall Sahlins, *Historical Metaphors and Mythical Realities* (Ann Arbor: University of Michigan Press, 1981).

Having served with Bligh, Matthew Flinders was also aware – not least given the mutiny experienced by Bligh 14 years earlier – of the many problems inherent in successful long-distance oceanic exploration. In addition to preventing mutinies and meeting Admiralty orders, Flinders knew that commanding the perfect voyage meant being able to command the narrative of his voyage. Cook's status, as Flinders knew, was largely fuelled by the narratives of his voyages. And so, even before his voyage, Flinders made Willingham Franklin an offer:

> I would have you a literary man, and probably it may be hereafter in my power to give you a lift into notice … I shall have both pleasure and credit in introducing a coz. of your description. You must moreover understand, that this voyage of ours is to be written and published on our return, I am now engaged in writing a rough account, but authorship sits awkward upon me, I am diffident of appearing before the public, unburnished by an abler hand. What say you? Will you give me your assistance, if on my return a narration of our voyage should be called for from me? If the voyage should be well executed and well told afterward, I shall have some credit to spare to deserving friends. If the door now opened suits your taste and you will enter it, prepare yourself for the undertaking.[48]

This was the self-publicising notice of a man who saw in exploration immense possibilities for improvement for himself and those associated with him in this voyage, but who knew, too, that with others, he had to manage in print the records of his achievements.[49]

Matthew Flinders did not achieve the authoritative authorship, or celebratory status, that would have made his trip successful after the event. Following its heavy leaking which was deemed irreparable, HMS *Investigator* was abandoned at Sydney in June 1803. Unable to find a suitable replacement, Flinders decided to return to England as a passenger on HMS *Porpoise* – on which he exerted enough official power to require of its captain that space be found for Flinders' collection of plants and papers. A little over a week after boarding, however, the *Porpoise* ran aground on a reef off Australia's north-east coast. Returning to Sydney in the ship's cutter, Flinders took command of a further ship, HMS *Cumberland*, soon to return to England. This ship, too, was soon in need of repairs, and so Flinders put in at French-controlled Mauritius in December 1803. This decision proved unfortunate. With Britain at war with France from May 1803, the French governor imprisoned Flinders, keeping him captive until an English blockade secured his release in June 1810.

---

48 State Library of New South Wales, Safe 1/55, Matthew Flinders' papers, Private letters, Volume 1, 'Matthew Flinders to Willingham Franklin, 1801'.

49 These issues are a central theme of Innes Keighren, Charles W.J. Withers and Bill Bell, *Travels into Print: Exploration, Writing, and Publishing with John Murray, 1773–1859* (Chicago: University of Chicago Press, 2015).

Matthew Flinders understood that absence from London meant that the promise of riches and the opportunity to oversee publication of his maps and the narrative of his discoveries would slip away from him. In contrast, astronomers such as William Wales and John Crosley were employed for years on their return from voyages to prepare the charts for publication. Flinders' narrative was mutable. He could, however, communicate with people in England via letter. Samuel Flinders had managed to return to England, and quickly became one of his brother's principal interlocutors. In his letters to Samuel Ward Flinders and in petitions sent to the Board of Longitude, Matthew Flinders was in turn concerned to have his brother acknowledged as astronomer and to be employed by the Board of Longitude to reduce and compute the astronomical observations made on the voyage which would, he hoped, be published by the Board of Longitude. Being invested even retrospectively with the authority of having acted as an astronomer would improve Samuel's standing at the same time as the public account of the voyage was in production. Matthew Flinders' 'perfect voyage' – far from perfect in reality – thus needed to be done through his representatives. Having a brother acknowledged as astronomer on the voyage would be more than simply valuable in testifying to Matthew Flinders' account of the voyage. It would also be helpful in publishing the results of the voyage – to help establish authorship of *his* voyage. As he said to Willingham Franklin: 'Seamanship and authorship make too great an angle with each other; the further a man advances upon one line the further distant he becomes from any point on the other'.[50]

For their part, the Board of Longitude and the Admiralty were greatly concerned with the relationship between 'seamanship' and 'authorship' and for the reasons we have seen. Both organisations were well aware of the problems that could ensue around competing claims by rival officers to have been the 'authors' of the charts. William Bligh, for example, after serving as sailing master on HMS *Resolution* on Cook's final voyage, had launched a bitter attack on Lieutenant Henry Roberts. As James Mumford has pointed out, on the title page of his personal copy of the accounts Bligh wrote: 'None of the Maps and Charts in this publication are from the original drawings of Lieut. Henry Roberts, he did no more than copy the original ones from Captain Cook who besides myself was the only person that surveyed & laid the Coast down, in the Resolution. Every Plan & Chart from C. Cook's death are exact Copies of my works'.[51] Authorship was always a matter of authenticity.

Following Vancouver's expedition in the 1790s, the Board was forced to confront a further instance of competing claims for reward. In 1797, Vancouver's aristocratic midshipman, Joseph Whidbey, petitioned the Board of Longitude claiming that he, and not Vancouver, had acted as the 'astronomer' with regard

50   State Library of New South Wales, Safe 1/55, Matthew Flinders' papers, Private letters, Volume 1, 'Matthew Flinders to Willingham Franklin, 1801'.

51   James Kenneth Munford, 'Preface' to *John Ledyard's Journal of Captain Cook's Last Voyage* (Oregon: University of Oregon Press, 1963).

to the voyage's observations.[52] Whidbey's petition centred on his claims that he had made the majority of observations and looked after the Board's instruments. Vancouver countered Whidbey's claims. As well as the Board being unable to come to a decision on this point – they did decide not to grant Whidbey any extra money – the affair combined with several other circumstances to cast doubts upon Vancouver's character. Vancouver was even attacked in the street by a group including Whidbey, in an event that quickly came to be referred to as the 'Caneing in Conduit Street'.[53]

The fact that officers in the Navy performed their astronomical and navigational tasks by relying on collaboration with others was not an issue for the Board of Longitude. What was at stake was the ability of different parties to assert control over the recording of the astronomical observations and the construction of the charts. The Board and its assigned astronomers understood that timekeepers and other instruments were liable to be lost, or to need regular repair. This itself was not damaging. What was more damaging for a London-based centre of control like the Board was its inability to salvage order from the potential disorder of unwritten, or illegible, accounts from the people it had sent out. That is why Maskelyne sent out his astronomers with clear notes of guidance. As he reminded them:

> You are to take particular care, that all your nautical and astronomical observations, and trigonometrical operations, whether made on ship-board, or on shore, with their results, be kept in a clear, distinct, and regular manner, in a book or books prepared for the purpose; and that they be written there in, with all their circumstances, immediately after they shall be made, or as soon after as they can be conveniently transcribed therein, from the loose papers or memorandum books, in which they may have been first entered; which book or books are to be always open for the inspection and use of the Commander & Master of the Vessel; and you are to send to the Astronomer Royal, by every safe conveyance which may offer, the results of your several observations, and also the principal observations themselves.[54]

Conclusion

Maskelyne and his Board of Longitude colleagues thought it important to send astronomers on voyages because it was thought they could be trusted to produce and to keep accurate accounts. Accurate accounting was based on mathematical and

52   CUL RGO 14/12, 'Letter from Joseph Whidbey to Board of Longitude claiming payment, February, 1797', 480.

53   Alison Gifford, *Captain Vancouver: A Portrait of His Life* (London: Allen Lane, 1986).

54   CUL RGO 4/185, 'Instructions for John Crosley to go on a voyage to New Holland on board HMS "Investigator" on a scientific expedition, 7 March 1801', 4.

astronomical skill. To demonstrate this, the appointed astronomers had to follow instructions that, in turn, represented and embodied experience at Greenwich and acquaintance with Nevil Maskelyne. These facts have implications for how we understand the relationship between astronomers as representatives of the Board of Longitude and officers in the Royal Navy. They also have implications for how we understand the role of those instruments, particularly timekeepers, carried on voyages during the Enlightenment. Instruments such as timekeepers could not work as stand-alone devices, nor was it desired that they should. Similarly, it was understood that the events which went to make up voyages could be prone to disintegration when attempts were made to reconstruct them upon the navigators' and the instruments' return. The solution to this, so the Board of Longitude reckoned, was to transform the practices of the Royal Navy by making its officers report, consistently and reliably. Reporting was a form of discipline. The relationship between instruments and timekeepers is therefore nuanced and mutual: because timekeeping instruments were not always reliable, one solution was to make officers themselves act like clocks.[55]

Marshall Sahlins has argued against historical and anthropological interpretations that have perpetuated 'the idea that the global expansion of Western capitalism, or the World-System so-called, has made the colonized and "peripheral" peoples the passive objects of their own history and not its authors, and through tributary economic relations has turned their cultures likewise into adulterated goods'.[56] It might be said that the relationship between the timekeeper and 'the ship' has a similar history. Oscillating between two deterministic poles, the late eighteenth- and early nineteenth-century ship has been characterised either as a non-participative courier of goods and knowledge, or as a homogenous spearhead of uniform metropolitan culture and power, capable of transforming the Atlantic and the Pacific peoples whose contact they engineered. Portrayed as a simple single 'thing' – of people, technologies, cultures, and histories – 'the ship' has come to be denied active agency as a site despite very specific and contingent relations occurring there, and its obvious role in linking peoples and objects across time and space. This chapter has shown that an understanding of the ways in which metropolitan authorities attempted to order their crews and their technologies, and how these had to work continuously to orient themselves back from whence they

---

55   On this point for a later period, see Randolph Cock, 'Scientific Servicemen in the Royal Navy and the Professionalisation of Science, 1816–1855', in *Science and Beliefs: From Natural Philosophy to Natural Science, 1700–1900* (ed.) David Knight and Matthew Eddy (Aldershot: Ashgate, 2005), 95–111, and Charles W.J. Withers, 'Science, Scientific Instruments and Questions of Method in Nineteenth-Century British Geography', *Transactions of the Institute of British Geographers* 38 (2013): 167–80.

56   Marshall Sahlins, 'Cosmologies of Capitalism: The Trans-Pacific Sector of "The World System"', in Nicholas B. Dirks, Geoff Eley and Sherry B. Ortner (eds) *Culture/ Power/History: A Reader in Contemporary Social Theory* (Princeton: Princeton University Press, 1994), 412.

came, sheds light on the problems brought about when an ensemble of people, technologies, and information moves together. Order and relationships cannot be assumed. They had to be reworked against and around ideas of exchange in which the role played by timekeepers and other astronomical instruments was determined by the social relationships in which they functioned. We should certainly not take for granted the notion of stable 'core' with effective products and agents acting upon an unstable periphery. This may have been the ambition of people who were attempting to transform the customs and practices of officers in the Royal Navy. But because these ambitions were differently realised, in understanding the nature and consequences of exploration it is vital that we understand the technologies and the instructive rhetoric driving that exploration.

Chapter 3

North by Northwest? Experimental Instruments and Instruments of Experiment

Richard Dunn

Nineteenth-century voyages of exploration to the Arctic had a range of purposes that included searching for new routes and exploitable resources, improving geographical knowledge, and investigating terrestrial phenomena. These required a range of instruments to measure different phenomena, yet the idea that these instruments could be seen as stable and trustworthy was never a given. Rather, the instruments were themselves often the subject of trial, experimental investigation and scrutiny, as were the practices of the observers using them.

This chapter looks at that most unreliable of instruments, the magnetic compass and takes as its case study the 1818 Arctic voyage of the *Isabella* and *Alexander* under the command of Captain John Ross. Long crucial to navigation, compasses on the 1818 voyage operated both as tools and subjects of investigation. As the 1818 voyage showed, these wayward instruments required considerable attention to be made to agree 'tolerably well', whether for routine navigation or scientific investigation. This proved equally true of the officers. Trust had to be earned and credibility crafted both through observational practice and the rhetoric of reporting.

To focus on the 1818 voyage opens up a range of issues. It reveals some of the challenges of establishing trust in an instrument long known to be untrustworthy, yet one which was indispensable for investigating magnetic phenomena in environments that rendered it even more unstable. It also allows one to explore the sometimes ambiguous status of instruments as simultaneously tools and subject of investigation, sometimes switching roles from experimental to authoritative and back again according to the particular situation and needs of the observer. The 1818 voyage offers rich evidence with which to explore these issues because several new designs of compass for both navigational and investigative purposes (of which four will be the focus of this chapter) were taken on the two ships and were tested and used for magnetic research in conditions that the observers found to be extremely challenging. These included theoretical challenges, in that magnetism was as yet poorly understood, and practical ones in a part of the world that was subject to extreme magnetic influences. What emerges is an account of some of the ways in which observers attempt to produce credibility and tolerable agreement in and through instruments in exacting conditions.

Magnetic Knowledge

Terrestrial magnetism and the workings of the magnetic compass were poorly understood in the early nineteenth century, despite centuries of using compasses at sea. In his historical survey of the 200 years to 1800, for example, Jonkers notes that, 'the geomagnetic field was even less well understood than at present. At sea it was both friend and foe, guiding mariners, but also confounding them by constantly changing appearance over the years. Therefore, a vigil had to be kept through frequent observation, stored in the ship's navigational logbook, and allowance made for needle deflection from true north in steering, dead-reckoning, and charting'.[1] As natural philosophers, hydrographers, and navigators attempted to analyse and understand the underlying patterns, local and global, more than 100 theories of the Earth's magnetic field were set down in the period between 1500 and 1800. Each differed in the number, position and motions of the magnetic poles postulated, but none seemed to describe the vagaries of the real world with any success.[2] As Jonkers has explained, even today, when huge historical datasets can be assembled and analysed, there is still no comprehensive theory that can explain and predict the changing patterns of terrestrial magnetism.[3]

The challenges presented by this acknowledged ignorance were equally conspicuous in the 1810s, but they also offered opportunities. As William Parry noted in relation to the 1818 Arctic expedition, 'there is no subject, connected with Science, which has proved so interesting as that little known and mysterious one of Magnetism'. He further remarked that, 'In this, as in many other instances, Science and Art go hand in hand; for this is a subject, which must eventually prove of equal interest to the philosopher, who speculates in his closet, and to the seaman, who traverses the oceans'.[4]

Seamen were well aware that the magnetic compass rarely pointed with any certainty towards true north. This was the result of two types of error: variation and deviation. Both were central to the investigations carried out on the 1818 voyage. The global phenomenon of magnetic variation – the angular difference between magnetic and true north – had been known for centuries, as had the facts that it changed over time and varied as one moved across the globe.[5] Although this might

---

1   A.R.T. Jonkers, *Earth's Magnetism in the Age of Sail* (Baltimore and London: Johns Hopkins University Press, 2003), 3.

2   For the background, see Jonkers, *Earth's Magnetism* and Patricia Fara, *Sympathetic Attractions: Magnetic Practices, Beliefs, and Symbolism in Eighteenth-Century England* (Princeton, NJ: Princeton University Press, 1996).

3   Jonkers, *Earth's Magnetism*, 3.

4   William Parry, 'Of the Phenomena in regard to Magnetism observed during the Voyage and particularly the Deviation of the Needle, on board the Alexander', The National Archives [hereafter TNA], ADM 55/4, 2, 62.

5   H.L. Hitchins and W.E. May, *From Lodestone to Gyro-compass* (London: Hutchinson, 1952), 39–40.

be problematic, it was correctable, or at least could be accounted for if regularly checked with observations of the Sun to determine true North. The more local and contingent deviation (often called 'local attraction' in this period) was the set of errors arising from the magnetic influence of a ship's and other nearby iron (or other magnetic material).[6] Although its effects were increasingly recognised, particularly as the use of iron and steel became more common for shipbuilding and ships' equipment, it is a more complex phenomenon than variation. The nature and causes of deviation were still unknown in 1818.[7]

The 1818 Arctic Voyage

In that era of retrenchment that followed the end of the wars with France in 1815, expeditions to the Arctic regions offered a conspicuous arena for peacetime operations for Britain's and other navies. John Barrow, second secretary of the Admiralty was one of the main champions of these voyages.[8] National pride was at stake, he explained, with Russian expeditions threatening to dominate one of the great remaining unexplored territories: 'It would therefore have been something worse than indifference, if, in a reign which stands proudly pre-eminent for the spirit in which voyages of discovery have been conducted, England had quietly looked on, and suffered another nation to accomplish almost the only interesting discovery that remains to be made in geography, and one to which her old navigators were the first to open the way'.[9]

The 1818 expedition typified this attitude. Its primary aim was simple, John Ross recording that, 'two ships were to be sent out, to ascertain the existence or non-existence of a north-west passage' by heading around the extreme northeast coast of America and into the Bering Strait, which was thought to be the likeliest candidate for an opening to that long-sought seaway.[10] Ross's instructions defined a secondary purpose: to improve geographic and hydrographic knowledge of the region. Given the evident peculiarities of the Earth's magnetic field and its strange effects on ships' compasses in the Arctic regions, this included magnetic researches in far northern waters.

---

6 According to Parry, 'Of the Phenomena', 14, it was John Ross who first used the term 'deviation' to refer to this local attraction.

7 A.E. Fanning, *Steady as She Goes: A History of the Compass Department of the Admiralty* (London: HMSO, 1986), 421–6; Alan Gurney, *Compass* (London: Norton & Co., 2004), 277–9; Hitchins and May, *From Lodestone to Gyro-compass*, 52–71.

8 Trevor Levere, *Science and the Canadian Arctic: A Century of Exploration 1818–1918* (Cambridge: Cambridge University Press, 1993), 41–4.

9 John Barrow, *A Chronological History of the Voyages into the Arctic Regions* (London: John Murray, 1818), 364–5.

10 John Ross, *A Voyage of Discovery Made under the Orders of the Admiralty, in His Majesty's Ships* Isabella *and* Alexander, *for the Purpose of Exploring Baffin's Bay, and Inquiring into the Possibility of a North-West Passage* (London: John Murray, 1819), iv.

To help with the investigations, Ross was informed that the crew would include, 'Captain Sabine, of the Royal Artillery, who is represented to us as a gentleman well skilled in astronomy, natural history, and various branches of knowledge, to assist you in making such observations as may tend to the improvement of geography and navigation, and the advancement of science in general. Amongst other subjects of scientific enquiry, you will particularly direct your attention to the variation and inclination of the magnetic needle, and the intensity of the magnetic force'.[11] Although the expedition was unquestionably a Royal Navy affair, Sabine had a background in the armed forces, the patronage of John Barrow and a recommendation from the Royal Society, which had helped define the voyage's scientific purposes. These smoothed the way for his participation, although his status as an outsider would resurface in the controversies that followed in the voyage's wake.[12]

The ships allocated and converted for the tough Arctic conditions were the *Isabella* (with a complement of 57 on board) under the command of John Ross and the *Alexander* (with 37 on board) under the command of William Parry. They were fully equipped with a range of instruments and equipment suitable for the proposed investigations. As Barrow boasted, no other polar expedition had 'been fitted out on so extensive a scale or so completely equipped in every respect'.[13]

In many ways, however, the voyage proved unsuccessful. After his return, Ross was criticised for his failure to penetrate Lancaster Sound in his search for a Northwest Passage, having turned the ships around after sighting what he thought were mountains which, mistakenly in hindsight, he took to indicate that the sound was closed off by land. Barrow wrote a scathing review of Ross's published account, from which the Admiralty had already withdrawn official support before publication, and Sabine offered further criticism in his 1819 book on the Ross voyage and his involvement in it.[14] Sabine aimed his attack not only at the allegedly hasty decision to turn back from Lancaster Sound, but also, and perhaps more heatedly, at Ross's scientific conduct, including Ross's observing practices and supposed misappropriation of the results of others. Ross, who had also effectively been chastised by an Admiralty enquiry, countered in print that, as commanding officer, he had full rights to all results and that he 'should not value

---

11   Ibid., 9–10.

12   Levere, *Science and the Canadian Arctic*, 46–7, 52; Gregory A. Good, 'Sabine, Sir Edward (1788–1883)', *Oxford Dictionary of National Biography* (Oxford: Oxford University Press, 2004; online edn, January 2011) (http://www.oxforddnb.com/view/arti cle/24436).

13   Barrow, *A Chronological History*, 364. In the peace following the end of the wars with France, this was one of the first major expeditions of a new era for the Royal Navy.

14   Edward Sabine, *Remarks on the Account of the Late Voyage of Discovery to Baffin's Bay* (London: R. and A. Taylor for John Booth, 1819).

the opinion of a landsman' in matters of navigation.[15] These were questions on which there would be no agreement, tolerable or otherwise, and the voyage gained a certain notoriety (Figure 3.1).[16] Leaving these broader issues aside, however, the discussion for the remainder of this chapter will concentrate on the deployment of compasses on the voyage, drawing principally from Ross's publication, Parry's manuscript notes, and Sabine's published papers on magnetism.

The Compasses

According to Ross, the voyage's impressive list of instruments included seven different types of magnetic compass (in addition to those already supplied to the ships for normal seafaring purposes), together with instruments such as dipping needles, for investigating the Earth's magnetic field.[17] Of the seven compasses that Ross identifies, one can think of them in two groups: azimuth compasses for taking bearings of observed objects; and steering compasses, used primarily for holding the ship's heading. Nonetheless, this distinction should not be seen as absolute, as there was crossover during the voyage. In thinking about the status of the different compasses and the trust they were accorded, it is worth highlighting four in particular.[18] Each was quite new in design – three having been designed within the previous decade – and each was to some extent on trial.

Kater's Azimuth Compass

The geodesist and metrologist Henry Kater noted in 1818 that, 'Of all instruments known, the common azimuth compass is perhaps the most defective'.[19] He had

---

15   John Ross, *An Explanation of Captain Sabine's Remarks on the Late Voyage of Discovery to Baffin's Bay* (London: John Murray, 1819), 15–16, 35.

16   Michael Bravo, 'Science and Discovery in the Admiralty Voyages to the Arctic Regions in Search of a North-West Passage (1818–25)', Unpublished PhD, University of Cambridge, 1992, 161–4; Levere, *Science and the Canadian Arctic*, 60–61; M. Ross, *Polar Pioneers: John Ross and James Clark Ross* (Montreal: McGill-Queen's University Press, 1994), 52–70; Elizabeth Baigent, 'Ross, Sir John (1777–1856)', *Oxford Dictionary of National Biography* (Oxford: Oxford University Press, 2004) (http://www.oxforddnb.com/view/article/24126).

17   Ross, *A Voyage of Discovery*, xvii–xix; Levere, *Science and the Canadian Arctic*, 50–55.

18   The 1818 expedition also tested Crow's steering and boat compasses and Burt's patent binnacle and compass.

19   Henry Kater, 'Instructions for the Adjustments and Use of the Instruments Furnished to the Northern Expeditions', *Quarterly Journal of Science and the Arts* 5 (1818): 202–30, quote from page 219. The instructions were printed for the Royal Society as *Instructions for the Adjustments and Use of the Instruments Intended for the Northern Expeditions* (London: W. Bulmer and Co., 1818).

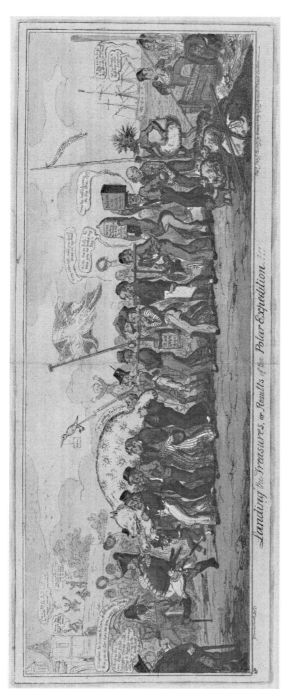

Landing the Treasures, or Results of the Polar Expedition.!!!

Figure 3.1 *Landing the Treasures, or Results of the Polar Expedition!!!,* **by George Cruikshank, published by George Humphrey, London, 1819**

Note: John Ross heads the procession, with Edward Sabine further back in army uniform, his bayonet spearing a gull.

Source: © National Maritime Museum, Greenwich [Photo ID: PX8511].

already proposed an improved design in 1811, which incorporated a mirror (or a prism in later models) so that the observer could read the compass's degree scale while also viewing a distant object through the sighting vane.[20] Ross's ships took three Kater compasses, two on the *Isabella*, one on the *Alexander*. They would prove to be crucial for the magnetic investigations.

Walker's Meridional or Azimuth Compass

This type of compass was originally designed by Ralph Walker in 1793 and submitted to the Board of Longitude as an instrument for finding longitude by magnetic variation (Figure 3.2). The design incorporated a sundial attachment on top, which allowed the user to determine true north from the Sun, which could be compared with magnetic north indicated by the compass card to find the variation. Although the Board of Longitude discounted it as a longitude-finding instrument, they gave Walker a reward of £200 for what was considered a promising design.

One of the compass's innovations concerned the mounting of the needle onto the card. Walker held that other compass needles were 'very improperly constructed' and noted that the steel used might be unevenly magnetised, so leading to an irregular response to the Earth's magnetic field. To combat this, he mounted the needle to the underside of the card along its thin edge rather than flat, as in other compasses. He anticipated that the needle would align more correctly with the Earth's magnetic field as there was less width across which its magnetic response might vary.[21] By 1818, the Royal Navy had bought a significant number of Walker's compasses, although expense and ease of use did sometimes limit their issue.[22] They were, however, largely being deployed as reliable, albeit high-end, azimuth compasses that could be used to provide the benchmark observations against which to assess other models.[23] One of these was taken aboard the *Isabella*.

---

20   A patent for Kater's design was taken out by Charles Schmalcalder in 1812 (no. 3545). A prototype of Kater's compass is in the Powerhouse Museum, Sydney, museum no. H9918.

21   Ralph Walker, *A Treatise on Magnetism, with a Description and Explanation of a Meridional and Azimuth Compass* (London: R. Hindmarsh, 1794), 50–52.

22   W.E. May, 'The Gentleman of Jamaica', *Mariner's Mirror* 73 (1987): 149–65; Board of Longitude, confirmed minutes, 6 June 1795, Cambridge University Library [hereafter CUL], RGO 14/6, 246–7 (reward of £200 agreed); see also Matthew Flinders, letter to John Croker, 26 April 1812, TNA ADM 1/1809, no. F89 (successful deployment of a Walker compass); Sir Charles Baxton, order, 26 October 1795, TNA ADM 106/2511, no. 238 (limiting issue to those qualified; rescinded in 1815); Navy Board, warrant, 10 August 1819, TNA ADM 106/2530, no. 441 (limiting issue due to large numbers of 'ordinary' azimuth compasses in store; rescinded in 1828).

23   For example, see Lt. Henry Reneau, 'Report on Jennings' Insulated Compass', 1 August 1818, TNA ADM 1/630, describing trials in which Walker's compass was used for azimuth and amplitude measurements for comparison with Jennings's and other compasses.

**Figure 3.2   Meridional and azimuth compass, by Ralph Walker,
               London, after 1793**

Source: © National Maritime Museum, Greenwich [Photo ID: L4514-001].

Jennings's Insulating Compass

The 'Insulating Compass' (Figure 3.3) was designed by Henry Constantine
Jennings, an idiosyncratic and irascible inventor of a rich miscellany of schemes.
Jennings first contacted the Admiralty with 'a method of making a Compass in
such a manner as always to point <u>Due</u> North, without any Variation' in 1815. As
well as producing an accurate compass, he claimed, it would allow mariners to
determine latitude and longitude when used with the appropriate tables.[24] Jennings
also submitted his proposal to the Navy Board and to the Board of Longitude,
which was at this time still considering and rewarding improvements in navigation
at sea.[25]

---

24   Henry Constantine Jennings, letter to the Board of Admiralty, 15 May 1815, TNA
ADM 1/4782, no. 141; Richard Dunn, 'The Impudent Mr Jennings and his Insulating
Compass', *Bulletin of the Scientific Instrument Society* 111 (2011): 6–9.

25   Henry Constantine Jennings, letter to the Secretary of the Board of Longitude,
August 1817, CUL RGO 14/31, ff. 204–205; Richard Dunn and Rebekah Higgitt, *Finding*

**Figure 3.3   Insulating Compass No. 6, by Jennings & Co., London,
about 1818, with the underside of its card showing
the curved iron 'shields'**

Source: © National Maritime Museum, Greenwich [Photo ID: L5875-001, L5875-004].

According to Jennings, his compass incorporated 'a means of guarding or protecting the magnetic needle … from all action arising from iron in its neighbourhood'. This was achieved by adding curved pieces of iron to the underside of the card (see Figure 3.3). These would act as 'guards against the passage of the magnetic fluid', with further shielding from specially treated iron filings in the sides of the brass compass box.[26] In a printed statement sent to the Board of Longitude, Jennings boasted that the compass's operation was founded upon nothing less than a 'new discovered Law of Nature', although the law was not disclosed.[27] Crucial to his claim was the possibility of separating the Earth's magnetic field and local attraction, with the compass responding to the former but not to the latter. In the 1810s, there was no unassailable reason to challenge this notion, since there was no agreed theoretical account of magnetism in either its global or local manifestations. So, while Jennings's claim to have discovered a new law of nature may seem somewhat grandiose, it did not undermine early nineteenth-century thinking about what remained an invisible and inexplicably

---

Longitude: How Ships, Clocks and Stars Helped Solve the Longitude Problem (Glasgow: Collins, 2014), chapter 6.

26   Henry Constantine Jennings, patent for 'An Improvement in the Mariners' Compass', 1818 no. 4259.

27   Henry Constantine Jennings, 'Mr. Jennings's Observations on the Improvement of the Mariner's Compass, by the Insulation of the Needle', 1817, CUL RGO 14/31, f. 203.

mysterious natural phenomenon. In the context of schemes for the longitude, Jennings's assertion also mirrored the strategies of other projectors in describing schemes said to be based on sound principles.[28] His words appear, therefore, to be a rhetorical claim rather than an assertion that new knowledge of nature had been or would be revealed through the application of instrumentation.

Although neither the Admiralty nor the Board of Longitude was inclined to respond to Jennings, the Navy agreed to test his compasses in 1818. As well as supplying a Jennings compass to Ross's expedition and to the concurrent expedition towards the North Pole under the command of John Buchan, naval tests of Jennings's compasses and new designs from other makers took place to the south of Ireland on the *Tonnant* between June and August 1818. Rear-Admiral Hallowell reported from these that Jennings's design did not seem to have 'any particular advantage … over the others', although, 'If the insulating or non-attractive Powers … can be established, it will be a great point gained'.[29] Lt Henry Reneau of the *Tonnant* was far more positive, reporting that 'a bright bar of Iron of half a pound weight' did not affect the compass (although it did derange other compasses). Reneau additionally observed how:

> that I have been an Eye witness to Mr Jennings's method of Insulating his Needle (on his shewing me the principle on which he did it) made me conclude, that if, it was only found to be equal in all respects to the others hitherto in use, that circumstance alone of its non-attractive quality, must render it more deserving than others, the attention of the Lords of the Admiralty, Navy Board, Merchants &c: … to me it appears possible, that a compass on Mr Jennings's principle <u>may be made</u> to resist the attraction of all the Iron contained in a Ship which must be the more desirable from the late successful introduction of Iron Knees, Cables &c.[30]

With Ross's expedition having departed by this time, the Jennings compass taken on the *Isabella* was still very much under trial.

George Alexander's Steering Compass

George Alexander, a watchmaker from Leith, took out a patent for a compass card with gimbals at its centre in 1813.[31] The design's advantages were said to include

---

28 See, for example, Richard Dunn, 'Scoping Longitude: Optical Designs for Navigation at Sea', in G. Strano et al. (eds) *From Earth-Bound to Satellite: Telescopes, Skills and Networks* (Leiden & Boston: Brill, 2011), 141–54, and see especially page 151.

29 Rear Admiral B. Hallowell, letter to the Admiralty, 2 August 1818, TNA ADM 1/630, item 148.

30 Reneau, 'Report on Jennings' Insulated Compass'.

31 George Alexander, patent for 'An Improved Mode of Suspending the Card of the Mariners Compass', 1813 no. 3646. An example of the card, by John Dickman, Leith, 1836,

the reduction of friction at the point of rotation, making the card more responsive and 'nimble' in its action. According to Alexander, it was also less vulnerable to 'unshipping' in high seas or when firing the guns. Like Jennings's insulated compass, it was trialled on the *Tonnant* and was judged to be very promising.[32]

Testing at Sea: The Compasses in Action

Soon after the voyage left Shetland, the difficulty of trusting the compasses became clear. As the two ships crossed the Atlantic, Parry noted substantial differences between their steering compasses. On 8 May 1818 (the fifth day after leaving Shetland), the *Isabella* steered NWbW by its steering compass, yet the *Alexander* was steering WNW by its compass.[33] The effects continued even after efforts were made to remove iron from near the *Isabella*'s binnacle by moving some of the guns and replacing the iron funnel of the gun-room store with a copper one.[34] Sabine later observed that as the *Alexander*'s compass was so close to the deck, it was affected by equipment such as ice anchors and ice saws, and that moving it a short distance greatly affected its readings.[35]

Things worsened as they headed north. The steering compasses of both ships, 'began to traverse so sluggishly, that it was necessary to shake the binnacles continuously … The same was observable in taking an azimuth with Walker's compass which, unless the observer constantly employed one hand in shaking it, would give no result near the truth'. Parry found that the metal rims of the cards in Burt's and Walker's compasses added considerable weight. They traversed freely if card was used instead of metal, but then 'the least damp made them bend, and therefore rendered them useless'.[36]

One unexpected problem arising from these difficulties was how to carry out routine tasks of seamanship. Parry found in Lancaster Sound that the wind direction according to the *Alexander*'s compass changed as the ship went onto a new tack. As he wrote:

---

is in the National Maritime Museum (museum no. ACO0924); see also *The Repertory of Arts, Manufactures, and Agriculture* 2nd Series XXII (1813): 255.

32   Reneau, 'Report on Jennings' Insulated Compass'.

33   It was standard practice to define headings according to points of the compass rather than in degrees. Normally 32 headings (11¼° apart) were specified. This was sufficient for routine navigation, given the vagaries of holding a precise course.

34   Parry, 'Of the Phenomena', 14.

35   Edward Sabine, 'On Irregularities Observed in the Direction of the Compass Needles of H.M.S. Isabella and Alexander, in Their Late Voyage of Discovery, and Caused by the Attraction of the Iron Contained in the Ships', *Philosophical Transactions of the Royal Society of London* 109 (1819): 112–22, see especially pages 114–15, 118.

36   Parry, 'Of the Phenomena', 10, 12.

It was amusing to see the dilemma into which the officers were thrown, by this unexpected phenomenon, to know <u>how to mark the wind</u> in the log book. In some instances, before the matter had been at all discussed, and the probable cause suggested, it was not unusual to make the wind shift, every time the helm was put down: and on one occasion, now on record in the log, a point is arbitrarily fixed upon, as the true Magnetic direction of the wind and the error left to be divided, in the best way it may, between both courses.[37]

Edward Sabine was more forthright about the implications as they headed north: 'The influence of the ship's iron on their compasses increasing, as the directive power of magnetism diminished, produced irregularities that rendered observations on board ship of little or no value towards a knowledge of the true variation'.[38] He produced a table to illustrate his point, which was based on readings from Walker's compass. For this investigation, Walker's compass was taken as a standard, albeit one to demonstrate the general unreliability of magnetic instruments in this peculiar environment. Shipboard observations were worthless in high latitudes, concluded Sabine, a problem that could not be remedied without a better understanding of a ship's magnetic influences on its instruments. On that point, at least, he agreed more than tolerably with Captain Ross.

What these problems threw up were questions of trust in the expedition's reported results. This was a matter of establishing the credibility of both the observers and of the instruments and procedures they were using.[39] The two went hand in hand, and because this was so, the deranging effects of the Arctic environment might call both into question. It was necessary, therefore, to enact and subsequently describe procedures that might secure credibility. This can be seen in Sabine's second published paper, which discussed attempts to render the 'standard' compass (the steering compass taken as the standard for recording headings for routine navigation) more faithful by trying to identify a magnetically isolated position. On the *Isabella*, they placed the compass amidships on a cross beam 9 or 10 feet above the deck. The *Alexander* had no cross beam, so the compass was placed on a box of sand on the companion, 5 to 6 feet above the deck.[40] Nonetheless, Sabine asserted that until local attraction was better understood, even 'the azimuth

---

37   Ibid., 32.

38   Edward Sabine, 'Observations on the Dip and Variation of the Magnetic Needle, and on the Intensity of the Magnetic Force; Made during the Late Voyage in Search of a North West Passage', *Philosophical Transactions of the Royal Society of London* 109 (1819): 132–44, quote from page 142, table at page 144; see also Bravo, 'Science and Discovery', 131, 154.

39   For issues of trust in individuals and in their spoken and written claims, see Steven Shapin, *A Social History of Truth* (Chicago and London: University of Chicago Press, 1994). For a discussion of the kinds of knowledge produced by instruments, see Davis Baird, *Thing Knowledge: A Philosophy of Scientific Instruments* (Berkeley and London: University of California Press, 2004).

40   Sabine, 'On Irregularities', 115–16.

compass was an imperfect instrument, and only to be relied on within certain undefined and variable limits'.[41] Not even the most precise of the compasses, he felt, warranted unquestioning trust in such a deranging environment.

Like Sabine, Parry noted that 'on board the Alexander, <u>a few inches</u>' alteration was sufficient to produce a very sensible difference in the observed Variation'.[42] The ships and their equipment were interacting with the compasses in ways that made the compasses untrustworthy. If one follows Sorrenson's line that a ship is a scientific instrument, this evidence would suggest that it was now becoming an instrument that was interfering with the others being deployed.[43] One obvious solution was to move observations well away from the ships to eliminate their deranging effects. As Parry wrote, 'I think we never considered it safe to take an azimuth or bearing upon the ice, within 150 yards of the ships, and, if possible, always removed to a still greater distance than this. For the same reason, the boat in which we landed upon the ice, was always sent to a sufficient distance, while we were observing'.[44] Ironically, therefore, further isolation became a desideratum in one of the most isolated places imaginable. This was crucial for ensuring that the reported results carried some credibility for readers in less unusual environments.

Despite these ongoing problems, Ross felt that the investigations of variation and deviation met with some success. 'Every possible opportunity was embraced during the voyage of taking observations, and of making all the necessary experiments and comparisons', he asserted.[45] Parry was equally optimistic: 'The opportunities ... offered ... of collecting facts on this subject [deviation], are such as never before occurred, in the annals of the World'.[46]

The first investigations into deviation involved comparing compass readings from a magnetically isolated location on the ice with readings using the same compass from a ship. These were performed in Baffin Bay by erecting a flagstaff on an island and taking the bearing, with a Kater's azimuth compass, from the flagstaff to a distinctive patch of snow on a distant 'sugar-loaf Hill'. The ships then lined up the two targets while sailing on as many different headings as possible, with azimuth bearings to the flagstaff taken with Kater compasses from two different positions on the *Alexander*. This experiment required the comparison of

---

41   Ibid., 113.

42   Parry, 'Of the Phenomena', 54.

43   Richard Sorrenson, 'The Ship as a Scientific Instrument in the Eighteenth Century', *Osiris* 11 (1996): 221–36.

44   Parry, 'Of the Phenomena', 30.

45   John Ross, 'On the Variation of the Compass and Deviation of the Magnetic Needle', in *A Voyage of Discovery*: Appendix I, quote from page viii. Ross sent handwritten copies of this report to the Royal Society and the Board of Longitude; the latter is at CUL RGO 14/42, ff. 319r–326r.

46   Parry, 'Of the Phenomena', 8.

readings from the Kater compasses with those from the ships' steering compasses in order, therefore, to produce a table of deviations for different headings.[47]

Parry noted some problems, however, in particular that of holding the ship on a specific heading, 'even in smooth water'. This was a challenge of seamanship – keeping the ship on a consistent course – and of trusting the compass to give a steady reading of the heading chosen. Two issues were crucial, he felt: the fact that the differences between observed bearings might be only a degree or less and thus difficult to discern; and that the results might be marred by the sluggishness of the compasses.[48] In his opinion, these difficulties invalidated the experiment. For similar experiments from the *Isabella*, Ross added that, 'observations were made … with Walker's, Alexander's, Jenning's [sic], and Burt's compasses; all of them agreeing. But Jenning's and Burt's did not traverse sufficiently quick to obtain all the results with them. The four compasses used, were always kept in the same stations, where they were found to agree with those in the binnacles'.[49] In this experiment, then, the compasses occupied two roles, simultaneously as part of the investigation of deviation and as equipment under test. An obvious problem, then, was that these investigations were falling foul of what has become known as 'experimenter's regress' – the problem of knowing whether an experimental result is correct unless you know what it should be, in particular when operating with new or experimental equipment – in that Ross appears to have been using instruments supposedly under test simultaneously to produce and/or verify results.[50] Such negotiations were, however, familiar to those using all sorts of instruments, also including pendulums and chronometers, on nineteenth-century expeditions.[51] Part of the observer's skill lay in credibly negotiating the apparent contradiction.

In another experiment, 'conducted with very great caution, and . . . under every favourable circumstance that can possibly occur on board a ship', the *Alexander* was anchored near an ice-floe and a well-defined object was sighted to the South, 'by a Kater's compass, upon the ice'. A compass was placed on a plank forward of the mizenmast and 7 feet above the deck and was used by Ross to 'regulate the direction of [the *Alexander*'s] head'. The compass selected for Ross's use 'was previously found to agree tolerably with those in the binnacles – I say tolerably, because these latter did not always agree with each other nearer than a quarter-point'. The trust conferred was contingent and partial, but was bolstered by the 'very great caution' and removal of the instrument to the magnetically neutral

47   Ibid., 16–18, 64.

48   Ibid., 16–18.

49   Ross, 'On the Variation of the Compass', xviii.

50   On the 'experimenter's regress', see Harry Collins and Trevor Pinch, *The Golem. What You Should Know about Science* (Cambridge: Cambridge University Press, 1998 edn), 97–103.

51   For a discussion of similar issues with respect to pendulums, in which Sabine was a notable participant, see Sophie Waring, 'Thomas Young, the Board of Longitude and the Age of Reform', unpublished PhD, University of Cambridge, 2014, chapters 2 and 3.

platform of ice. With the nearby iron removed and the *Isabella* taken to a safe distance, the *Alexander* was then rotated using ropes through a series of headings as dictated by Ross, while Parry took azimuth observations to the target with a Kater compass placed on the companion, exactly amidships.[52]

What one begins to understand from these experiments is that different compasses on the two vessels were accorded different levels of trust and authority. Kater's azimuth compasses generally gained the highest level of trust because they appeared to give the most consistent and stable results, and were used as the benchmark for most (but not all) of the more rigorous investigations. Parry noted that the value of Kater's azimuth compass 'was such, that without it, not a single satisfactory observation for the Variation could perhaps have been obtained'.[53] Sabine agreed: 'the satisfactory results which have been obtained, even under such extreme circumstances with Captain KATER'S compasses, afford the best testimony of their excellence, and of the precision which may be expected from them in the ordinary course of observation'.[54] They were the best available, even if there remained a note of caution about their use in extreme magnetic environments. They also required some manipulation by the observer. Sabine found, for instance, that the Kater compasses were most effective when fitted to the gimbals designed for Alexander's steering compasses.[55]

Nonetheless, this was not a simple case of Kater's design being trusted to the exclusion of all others and in every circumstance. Parry's observations show that while on the *Alexander* the Kater compass was used for most of the variation readings, Walker's compass was used on some occasions.[56] Likewise, during the Atlantic crossing, Parry recorded how:

> Capt.[n] Sabine came on board the Alexander, at sea, for the express purpose of proving, if possible, by direct comparison, if any and what difference existed between the compasses on board the two ships. For this purpose, he brought with him a compass invented by Mr Jennings, and called by him the Insulating Compass … The result … appears to be highly in its favour, for when the compass in Burt's binnacle on board the Isabella shewed her head to be SW, Jennings's shewed SW½S: and when the Alexander's binnacle-compass shewed SW¼S, Jennings's shewed SW¼W: so that the Insulating Compass was a mean between the other two, and probably, therefore, indicated the true Magnetic Course.[57]

Here, Jennings's compass shifted from being a compass primarily under trial to one that could be used to interrogate and mediate the performance of the other

---

52 Parry, 'Of the Phenomena', 18–22; Sabine, 'On Irregularities', 117–18.
53 Parry, 'Of the Phenomena', 10.
54 Sabine, 'Observations on the Dip', 141.
55 Bravo, 'Science and Discovery', 131.
56 Parry, 'Of the Phenomena', 67–8.
57 Ibid., 8–10.

compasses, allowing the observers to draw meaningful conclusions about the effects of local attraction. Or rather, there was a circular move – the experiment validated the Jennings compass, allowing the investigators to draw tentative conclusions about the behaviour of the steering compasses. Again, this seems to have been part of a routine set of negotiations with instruments whose operation and performance was still, in principle, unproven.

Deviations and Disagreements

The findings from these trials and tests, whether from routine navigation or directed investigations, may be categorised as of two types. The first concerned the reliability of specific compasses (that is, testing the instruments). The second involved provisional conclusions about magnetic phenomena (using the compasses as investigative instruments).

Ross's published account included assessments of each of the new compass designs, underlining the fact that they were to be seen as experimental and unproven.[58] Agreeing with Parry and Sabine, Ross found Kater's azimuth compass excellent for measuring variation, although the device required careful steadying. He also thought it effective for investigating deviation. Walker's compass was 'certainly the best for azimuths when the ship has considerable motion; but its card being heavy, it ceased to traverse when the variation was 110°, and the dip 86°'; in other words, when the horizontal force of the Earth's magnetic field was weak.[59] The compass warranted some trust, but it had limitations.

The Jennings insulating compass, which had become a kind of standard in the early investigation of the steering compasses, 'answered the purpose for which it was intended, and completely obviated the effect of local attraction; but the card being heavy, and the needle short and not very powerfully magnetized, it ceased to act when the variation was great'.[60] It could only be trusted to a limited extent. George Fisher, astronomer on Buchan's voyage towards the North Pole, also made tests of Jennings's compass, finding that it 'certainly appears to resist the action of iron not magnetical', although it was 'not less affected than the others, and continued so throughout the voyage'.[61] Despite these tentative conclusions and some much more effusive recommendations from other seafarers, later history would dismiss Jennings's invention. In 1818 it seemed to warrant trust and further

---

58   John Ross, 'Report on Compasses, Instruments, &c.', in *A Voyage of Discovery*: Appendix, cxxiv–cxxvi.

59   Ibid., cxxiv.

60   Ibid., cxxv.

61   George Fisher, 'On the Variation of the Compass', *Quarterly Journal of Science, Literature and the Arts* IX (1820): 81–106, quote from page 81.

encouragement; by the middle of the century, advances in theories of magnetism would allow it to be dismissed as entirely implausible.[62]

Ross found that Alexander's steering compass was 'decidedly superior to all the others … well adapted either for boats or ships, and if fitted as an azimuth compass cannot fail to excel, particularly when the ship has much motion: those we had on board the Isabella and Alexander traversed when all others ceased to act'.[63] The new design offered hope that it was a compass that might span both navigational and observational/investigative roles (although it does not seem to have been adopted by the Navy in either capacity).

The results that the compasses produced in investigations of magnetic deviation reveal the continuing uncertainty arising from an inherently untrustworthy, albeit crucial, instrument. In speculating on the causes of deviation, which so affected the investigations of variation, Parry suggested that the local attraction from a ship's iron seemed to follow the laws of the magnet. A ship was like an artificial magnet with many poles, he wrote, postulating that terrestrial location was significant and that the relative effects were likely to increase with latitude.[64] His investigations also convinced him that previous hypotheses, particularly those put forward by Matthew Flinders, were flawed. But Parry remained cautious, admitting that:

> it may justly be complained that little service is done to Navigation, by merely pointing out the inadequacy of the former rule, unless some more effectual one can be substituted for it. It must be confessed, however, that the few facts which have yet been collected do not appear to offer a hope of any rule being at present established, which is likely to prove of general utility. But the subject, it must be remembered, is yet in its infancy; for, in its present more extended view, the late Voyage may be said to have given birth to it.

It was a matter that would require more data-gathering with the right instruments. He did, however, suggest some lines of enquiry that might pin down how deviation altered on different headings and the extent to which it displayed symmetries.[65] John Ross was bolder and offered some more specific conclusions:

1. Every ship has an individual attraction.
2. Because attraction is individual, all previous rules must be considered unreliable.
3. Deviation varies according to where the compass is in a ship.
4. Deviation varies according to the ship's heading.
5. Deviation appears to be affected by heat and cold and by variations in the humidity and density of the atmosphere.

---

62   Dunn, 'The Impudent Mr Jennings', *passim*.
63   Ross, 'Report on Compasses', cxxv.
64   Parry, 'Of the Phenomena', 52.
65   Ibid., 56–60.

6. Wind direction appears to have 'an effect in disturbing the regularity of the deviation'.
7. Magnetic dip appears to have an effect.
8. There appears to be a relationship between deviation, variation and dip.[66]

Later investigations and theoretical work would dismiss some of Ross's suggestions, as well as Jennings's claim that one could insulate a compass from local attraction yet still have it respond to the Earth's field. In the 1810s, however, work with instruments that could be trusted to a limited degree only was still struggling to record and explain the phenomena. Establishing which compasses to trust, and in which circumstances, was no trivial matter in a geographical region prone to such deranging magnetic effects.

Conclusion

It is possible to make three general points from this individual case study. First, and perhaps most obviously, instruments such as compasses can and did have shifting roles: at one moment a quotidian tool for steering a ship or for surveying and charting; at another, an instrument of exploration used in the investigation of geophysical phenomena. Many of the compasses taken in 1818 were designed with one or other purpose in mind (for example, Kater's were intended as instruments of observation/investigation), although some were designed with an eye to both or, as in the case of Alexander's, seemed to offer hope that they could be made to fulfil both functions. This was not unique to compasses. There are other instances in which the tools of routine navigation, such as depth sounding equipment, have become instruments of investigation, filling both roles simultaneously.[67] In such contexts, the contingent practices of the observers clearly have a key role in defining or redefining the mode of operation, and the credibility in these different modes, of specific instruments. Observers' practices, and how they are subsequently described to remote readers, become crucial in this context, with careful operation and textual phrasing – Parry's 'very great caution', for example – conferring stability in the instrument and in the author, both real and assumed.[68]

Secondly, trust in instruments must be seen as contingent and specific rather than enduring. The reasons for trusting specific instruments may not always be clear to the later historian since they may relate to factors that included but were

---

66   Ross, 'On the Variation', xxx–xxxii.

67   On this point, see Sarah Louise Millar, 'Science at Sea: Soundings and Instrumental Knowledge in British Polar Expedition Narratives, *c*.1818–1848', *Journal of Historical Geography* 42 (2013): 77–87.

68   On these issues more generally, see Innes Keighren, Charles W.J. Withers and Bill Bell, *Travels into Print: Exploration, Writing, and Publishing with the House of Murray, 1773–1859* (Chicago: University of Chicago Press, 2015).

not restricted to their 'performance'. It is now difficult, for instance, to reconstruct the processes that accorded the Jennings insulating compass a significant degree of plausibility and some level of trust in 1818. For some observers, this did seem in part to derive from the compass's performance while on expedition, perhaps coupled with some anticipation of its supposed capabilities as a result of the faith and trust placed in Jennings's claims. For others, the trust they afforded instruments came from witnessing their inventor's explanation of how it was made and worked. Yet, for a while, this device appeared to show promise. Jennings's persistent lobbying ensured that his compasses went for trial on several ships; assessments of their performance reflected the anticipation he generated, but with the moderation of more local factors such as performance against (or in the investigation of) other compasses and tentative expectations regarding the phenomena under investigation. Kater's design had a quite different context. While Jennings was an unknown outsider, Kater was a well-established figure in British scientific circles by 1818, known to Sabine and an adviser on instrumental matters for expeditions.[69] His name conferred credibility and warranted specific efforts to make his instruments work in the field, whether by isolating them from the deranging effects of naval vessels or stabilising them with parts from other compasses.

Lastly, an instrument used for investigation may be simultaneously under test and an instrument that sets test benchmarks. This chapter has examined compasses, with a number of the compasses on the 1818 voyage operating at times in this dual role. Chronometers and other instruments on exploratory voyages might well repay investigation in similar ways. While there might appear to be a danger of an experimenter's regress in these situations, pragmatic negotiations around this problem were clearly familiar to nineteenth-century investigators, in particular those grappling with the contingent vagaries of research in the field.

---

69   Julian Holland, 'Kater, Henry (1777–1835)', *Oxford Dictionary of National Biography* (Oxford: Oxford University Press, 2004; online edn, October 2007) (http://www.oxforddnb.com/view/article/15186).

Chapter 4

Weather Instruments all at Sea: Meteorology and the Royal Navy in the Nineteenth Century

Simon Naylor

Over the last two decades historians and geographers of science have paid increasing attention to science in the field. For one, 'fieldwork has become the ideal type of knowledge', so much so that much work in science studies asks not 'about temporal priorities but about spatial coordination'.[1] This agenda has been pursued empirically through study of European exploration in the eighteenth and nineteenth centuries. One of the key problematics for historians and geographers has been how, exactly, science collaborated with state actors to extend European nations' 'spatial grip'. There have been three common empirical responses to this question: through the deployment of physical observatories; through fieldwork; and by means of ships. Studies of observatories – on mountains, on the edge of oceans, in the polar regions – are many, as are studies of the ephemeral fieldsite.[2] The ship has been seen to embody both of these types of scientific space. In his 1845 address to the British Association for the Advancement of Science (BAAS), the astronomer John Herschel referred to ships as 'itinerant observatories'.[3] Naval ships were deemed to be crewed by disciplined observers (the equivalent to observatories' 'obedient drudges'[4]), to run according to military discipline and were replete with the latest instruments. They were, then, no different to terrestrial physical observatories or laboratories, except for their mobility. Following Beaglehole, Sorrenson has argued that ships were more than floating laboratories;

---

1   Simon Schaffer, '"On Seeing Me Write": Inscription Devices in the South Seas', *Representations*, 97 (2007): 90–122, quote from page 91.

2   Henrika Kuklick and Robert Kohler (eds) *Science in the Field* (Chicago: University of Chicago Press, 1996); Michael Reidy, 'From the Oceans to the Mountains: Spatial Science in an Age of Empire', in J. Vetter (ed.) *Knowing Global Environments: New Historical Perspectives on the Field Sciences* (New Brunswick: Rutgers, 2011), 17–38.

3   Alison Winter, '"Compasses All Awry": The Iron Ship and the Ambiguities of Cultural Authority in Victorian Britain', *Victorian Studies* 38 (1994): 69–98, quote from page 75.

4   'Obedient drudges' was the term the Astronomer Royal, George Airy, used to describe the workers in the Royal Greenwich Observatory: Winter, '"Compasses All Awry"', 74.

ships were themselves instruments of geographical discovery – conferring authority on their user, leaving traces on maps, and providing 'superior, self-contained, and protected views of the landscapes' viewed from them.[5]

This chapter engages with debates about the ship as a site of scientific labour through an examination of the study of meteorology at sea in the period between the conclusion of the Napoleonic Wars, and the Conference on Maritime Meteorology, in London in 1874. It examines the roles played by individuals and institutions, guidebooks and regulations, in promoting a culture of instrumental meteorology onboard voyages of exploration, and on Royal Naval and Hydrographic Office survey ships.[6] Particular attention is paid to attempts to establish international standards for the study of meteorology at sea. The chapter illustrates how the British Admiralty was supportive of science in the nineteenth century in consenting to its ships being turned into floating meteorological observatories. The Admiralty did so to develop philosophical inquiry and to respond to more utilitarian concerns. Over the course of the nineteenth century an informal, even idiosyncratic, culture of meteorological inquiry was gradually formalised; uniform forms and meteorological instruments were introduced together with prescribed observations and practices. These, in turn, were authorised in Admiralty regulations and guidebooks. Voyages of exploration and the Hydrographic Office survey vessels were the experimental sites for this new culture, which was expanded to include all Royal Naval vessels, and, later, Britain's merchant marine.

Schaffer reminds us that the 'immutability' of scientific inscriptions made during voyages of exploration 'was a complicated and exhausting achievement'.[7] This was true for meteorological inscriptions, whether on a ship's log board, in a register, or on an instrument. In elucidating the adoption, use, and evaluation of meteorological instruments onboard British and other European and American ships, the chapter supports Schaffer's insistence that philosophical instruments existed in various 'states of repair'.[8] The term refers to the assumption that the demands of science routinely outstripped instruments' abilities and the humans that interacted with them. While I argue that the Admiralty was supportive of science, I also suggest that it was slow and conservative over the adoption of a new observational regime and the use of new instruments. Despite the adaption of meteorological instruments for life at sea, the conduct of sea trials, the issuance of

---

    5   Richard Sorrenson, 'The Ship as a Scientific Instrument in the Eighteenth Century', in Henrika Kuklick and Robert Kohler (eds) *Science in the Field* (Chicago: University of Chicago Press, 1996), 221–36, quote from page 222.

    6   Richard Dunn, '"Their Brains Over-Taxed": Ships, Instruments and Users', in Don Leggett and Richard Dunn (eds) *Re-inventing the Ship: Science, Technology and the Maritime World* (Aldershot: Ashgate, 2012), 131–55.

    7   Schaffer, '"On Seeing Me Write"', 107.

    8   Simon Schaffer, 'Easily Cracked: Scientific Instruments in States of Disrepair', *Isis* 102 (2011): 706–717.

guides to observation and regulations for their use, Britain's naval ships and those of other countries, were by no means itinerant observatories.

The chapter begins by summarising the changes that affected meteorological science and the British Navy in the early nineteenth century. It then examines the role of the Royal Society in the organization of several voyages of exploration in the 1820s and 1830s. It details the role played by Francis Beaufort in the promotion of the study of meteorology on board Royal Naval ships. The final section examines the aims and outcomes of two international conferences on maritime meteorology, in Brussels in 1853 and London in 1874.

Reforming Science and the Navy

Schaffer's claim that 'Managing states of disrepair is salient during scientific practices' periods of dislocation and reorganization, such as the later eighteenth- and early nineteenth-century scientific, industrial and political revolutions', is, I contend, applicable to the Admiralty and to meteorological science, in their periods of reform in the early nineteenth century.[9] The early 1800s witnessed the emergence of natural science out of natural philosophy – the development of 'a comprehensive science (a *physics*), which associated the separate branches of natural philosophy under a dynamic equilibrium model of natural order'.[10] Its geographical scope was ambitious: to chart the variation of physical and biological features on a global scale.[11] The physical sciences placed emphasis on data collection during the early nineteenth century, with an insistence upon trained observation, developments in written recording, and repetition of numerical measurement as a result of an increased reliance upon instrumentation'.[12] Measurement, quantification and mathematically stated laws were upheld as ideals for terrestrial physics in general, while the discipline of quantification shaped practice.[13]

---

9   Schaffer, 'Easily Cracked', 709.

10   Michael Dettelbach, 'Humboldtian Science', in Nicholas Jardine et al. (eds) *Cultures of Natural History* (Cambridge: Cambridge University Press, 1996), 287–304, quote from page 300, original emphasis.

11   Gregory Good, 'A Shift of View: Meteorology in John Herschel's Terrestrial Physics', in J.R. Fleming, V. Jankovic and D.R. Coen (eds) *Intimate Universality: Local and Global Themes in the History of Weather and Climate* (Sagamore Beach: Science History Publications, 2006), 35–68.

12   Charles W.J. Withers, 'Science, Scientific Instruments and Questions of Method in Nineteenth-Century British Geography', *Transactions of the Institute of British Geographers* 38 (2012): 167–79, quote from page 170. See also David Cahan (ed.) *From Natural Philosophy to the Sciences: Writing the History of Nineteenth-Century Science* (Chicago: Chicago University Press, 2003).

13   Good, 'A Shift of View', 36; Fabian Locher, 'The Observatory, The Land-Based Ship and the Crusades: Earth Sciences in European Context, 1830–50', *British Journal for the History of Science* 40 (2007): 491–504, quote from page 501; Ben Marsden and Crosbie

These reforms are clearly evident in the study of weather. Meteorology moved away from what Jankovic has termed the 'place-centred and curiosity-driven authority of meteoric reportage', to become 'a constellation of physico-chemical enquiry into the nature of atmospheric air and its planetary circulation'. The atmosphere was gradually re-conceived as a laboratory and the promotion of its instrumental measurement was justified in those terms.[14] Meteorology aimed to pursue this agenda on a global stage, employing trained observers and calibrated instruments, observing at fixed times of day using uniform methods and quantitative systems of notation, with a view to laying down 'the empirical laws of the atmosphere and perhaps extend them into a comprehensive meteorological theory'.[15] Edwards has suggested that meteorologists in the early-to-mid-nineteenth century employed instruments and methodical and simultaneous observations to effect a shift from a culture of voluntarist internationalism, 'based on an often temporary confluence of shared interests, to quasi-obligatory globalism based on a more permanent shared *infrastructure*'.[16] This was vital in studying localities with different weathers – such as between temperate northern Europe and its tropical colonies – and in the production of global climate maps, both of which 'required the fashioning of transnational and objective credibility supplied by instruments, standardized registers and regular observations'.[17]

The British Admiralty went through a period of reform in the 1810s and 1820s. In the aftermath of war with France, the Royal Navy experienced financial retrenchment and disarmament: many ships were decommissioned and thousands of enlisted men lost their jobs. Naval officers had greater political influence and so few of them were retired but perhaps 90 per cent of them found themselves unemployed and on the half pay list.[18] The First Secretary of the Admiralty, John Wilson Croker, defended the reduced Navy Estimates. The Second Secretary of the Admiralty, John Barrow, argued that the Navy's ships and personnel should be employed in global exploration, on the basis that 'exploration would increase scientific knowledge, that it would be a boon to national commerce, and above all that it would be a terrible blow to national pride if other countries

Smith, *Engineering Empires: A Cultural History of Technology in Nineteenth-Century Britain* (Basingstoke: Palgrave, 2005), 24.

14   Vladimir Jankovic, *Reading the Skies: A Cultural History of English Weather, 1650–1820* (Manchester: Manchester University Press, 2000), quotes from pages 164, 143, and 166 respectively.

15   Vladimir Jankovic, 'Ideological Crests versus Empirical Troughs: John Herschel's and William Radcliffe Birt's Research on Atmospheric Waves, 1843–50', *British Journal for the History of Sci*ence 31 (1998): 21–40, quote from page 28.

16   Paul N. Edwards, 'Meteorology as Infrastructural Globalism', *Osiris* 21 (2006): 229–50.

17   Jankovic, *Reading the Skies*, 158.

18   Christopher Lloyd, *Mr. Barrow of the Admiralty: A Life of Sir John Barrow* (London: Collins, 1970), 91–2.

should open up a globe over which Britain ruled supreme'.[19] The Royal Navy and the Admiralty Hydrographic Office made numerous contributions to science, including geographical exploration of the Northwest Passage, the Antarctic Ocean, and of Africa, such that 'in the first half of the nineteenth century the Navy was the principal governmental subsidizer of science'.[20] For naval officers interested in science, a position on one of these voyages of exploration was a choice appointment. These 'scientific servicemen' gradually took on much of the scientific work from civilians and many became Fellows of the Royal Society.[21]

Cultures of Instrumentation on Voyages of Exploration

Voyages of exploration in the late 1810s and 1820s served to establish standards for the conduct of physical scientific inquiry at sea, particularly in relation to the use of philosophical instruments onboard ship. The Royal Society had long offered advice to the Admiralty on the scientific aspects of its expeditions, viewed by government and the military as a 'state tool for consultation'.[22] The period from Ross's 1818 Arctic voyage to Foster's South Atlantic expedition in 1828 was a tumultuous one for the Society. Joseph Banks's reign as President of the Royal Society ended with his death in 1820. Successive presidents – Humphry Davy (1820–1827) and Davies Gilbert (1827–1830) – were caught up in wider contests over the character and direction of British science. Davy put the Royal Society on a course that aimed to satisfy both the remnants of Banks's 'Learned Empire' and the reformist intentions of the 'Cambridge Network'.[23] The changes experienced by the Society over this period were reflected in the composition and work of its committees. In the early years of Davy's presidency in particular, increased

19   Fergus Fleming, *Barrow's Boys: A Stirring Story of Daring, Fortitude and Outright Lunacy* (London: Granta, 1999), 11.

20   Alfred Friendly, *Beaufort of the Admiralty: The Life of Sir Francis Beaufort 1774–1857* (New York: Random House, 1977), 289.

21   David Philip Miller, 'The Revival of the Physical Sciences in Britain, 1815–1840', *Osiris* 2 (1986): 107–34; Randolph Cock, 'Scientific Servicemen in the Royal Navy and the Professionalisation of Science, 1816–55', in David M. Knight and Matthew D. Eddy (eds) *Science and Beliefs: From Natural Philosophy to Natural Science, 1700–1900* (Aldershot: Ashgate, 2005), 95–112.

22   Sophie Waring, 'The Board of Longitude and the Funding of Scientific Work: Negotiating Authority and Expertise in the Early Nineteenth Century', *Journal for Maritime Research* 16 (2014): 55–71, quote from page 57.

23   David Philip Miller, 'Between Hostile Camps: Sir Humphrey Davy's Presidency of the Royal Society of London, 1820–1827', *British Journal for the History of Science*, 16 (1983): 1–47; Roy M. MacLeod, 'Whigs and Savants: Reflections on the Reform Movement in the Royal Society, 1830–48', in Ian Inkster and Jack Morrell (eds) *Metropolis and Province: Science in British Culture, 1780–1850* (London: Hutchinson, 1983), 55–90.

use was made of scientific committees.[24] Over the course of the 1820s scientific reformers, such as John Herschel, Charles Babbage and Francis Baily, joined long-standing members like Thomas Young, Henry Kater, and William Hyde Wollaston, all taking a greater role in the running of these committees. Miller notes that members of the reform group 'increasingly dominated public discussion of the most important objects of research for scientific voyages'.[25] Herschel, in particular, 'maintained an ambition to make the surveying voyages commissioned by Barrow on behalf of the Admiralty more 'scientific'.[26] The changes effected in this period had a direct bearing on the advice that the Royal Society provided to exploring expeditions.

During the final years of Banks's presidency, William Thomas Brande, one of the Royal Society's two secretaries, wrote to Barrow to supply the Admiralty with a list of instruments that the Society recommended for use on the two 1818 expeditions then heading for the polar regions, to be led by John Ross and David Buchan respectively. These included compasses, barometers, magnetic instruments, bottom sampling and dredging equipment, chronometers, mercurial and sea thermometers, a Wollaston micrometer, artificial horizons, electrometers, hydrometers, and apparatus 'for ascertaining the quantity of air in water'.[27] Four laboratory tents were added to protect the instruments during observations to be made onshore, along with two transit instruments, four 'Small Altitude Instruments', a water sampler, and a tent for astronomical observations.[28]

In 1821, a 'Committee for suggesting Experiments and Observations to Mr Fisher, about to proceed to the Arctic Seas under the command of Capt. Parry' was established.[29] John Herschel, William Hyde Wollaston, and Charles Hatchett bolstered a core group made up of the President, the two secretaries – Brande and Taylor Combe – as well as Henry Kater and Thomas Young. The expedition astronomer, George Fisher, was invited to attend.[30] While the advice given to Ross in 1818 laid out in detail the instruments to be used on his expedition, that provided to Fisher was more direct in the scientific agenda to be pursued,

24   Maria Boas Hall, 'Public Science in Britain: The Role of the Royal Society', *Isis* 72 (1981): 627–9.

25   Miller, 'Between Hostile Camps', 34.

26   Waring, 'The Board of Longitude', 59.

27   Copy of letter from William T. Brande to John Barrow, 29 January 1818, Minutes of Committees of the Royal Society Appointed From Time to Time for Particular Purposes, CMB1/25, Archives of the Royal Society, London, 9.

28   'At a Committee for ascertaining the Length of the Seconds Pendulum', Minutes of Committees of the Royal Society Appointed From Time to Time for Particular Purposes, 26 March 1818, CMB1/25, Archives of the Royal Society, London, 14.

29   'Hints of Experiments to be made in the Arctic Expedition ... of 1821', Minutes of Committees of the Royal Society Appointed From Time to Time for Particular Purposes, 12 April 1821, CMB1/25, Archives of the Royal Society, London, 26–31.

30   G.W. Roberts, 'Magnetism and Chronometers: The Research of the Reverend George Fisher', *British Journal for the History of Science* 42 (2009): 57–72.

emphasising terrestrial physics. Twenty experiments were proposed. The majority focused on the effects of extreme cold on atmospheric chemistry, the behaviour of fluids (including mercury), and on humans, animals, food, and different metals. Of particular interest was the freezing point of pure mercury and of different amalgams of mercury and other metals. This was significant because of its effect on the performance of the thermometer and barometer.[31] Other questions related to the operation and effects of the Aurora Borealis, and the investigation of sea temperature at different depths.

The advice supplied to Captain Henry Foster's 1828 voyage on the HMS *Chanticleer* to the South Atlantic was more comprehensive still. At this committee Davies Gilbert (now President), Herschel, and Kater were joined by William Fitton, President of the Geological Society of London, Edward Sabine, artillery officer and expert in magnetism, and the Admiralty Hydrographer, Francis Beaufort. James Horsburgh, the East India Company Hydrographer, and Captains Parry and Foster, were present by invitation. In line with the interests of Herschel, Sabine, and Beaufort, the principal objects of Foster's expedition were defined as the investigation of physical astronomy, the determination of the figure of the earth, and the investigation of the law of the variation of gravity, along with inquiries into ocean currents, magnetism, the longitude of significant locations, natural history, and meteorology. The Committee noted that meteorological observations 'form a branch of inquiry of no small amount in this and all similar expeditions' and it recommended that 'regular observations of the Barometer, Thermometer, Hygrometer, and the direction and force of the wind should be daily made; and of the actinometer or other instruments proper for measuring the Solar and terrestrial variation, at favorable opportunities and at various levels'. The result, it was hoped, would be a better understanding of 'the probable former and future climate of different regions of the Earth[,] the permanence or variability of the Solar influence at different epochs, and the stability of the actual equilibrium of meteorological agents'.[32] In its findings, the voyage was judged a success and the results were later used by Royal Society reformers and members of the Astronomical Society to affirm the analytic importance of mathematics in accurate observation and experimental research.[33]

The advice given to Foster was dwarfed, however, by that supplied to James Clark Ross for his 1839 voyage to the Antarctic Ocean as part of the magnetic

---

31   Catherine Ward and Julian Dowdeswell, 'On the Meteorological Instruments and Observations Made during the 19th Century Exploration of the Canadian Northwest Passage', *Arctic, Antarctic, and Alpine Research* 38 (2006): 454–64, quote from page 455.

32   'At a meeting of the Committee for considering and resolving on the most advantageous objects to be attained by Capt'n Foster in the course of his intended scientific Voyage', Minutes of Committees of the Royal Society Appointed From Time to Time for Particular Purposes, 28 January 1828, CMB1/25, Archives of the Royal Society, London, 230.

33   Miller, 'The Revival of the Physical Sciences', 123.

crusade. The committee convened to advise on the expedition was chaired by Herschel and included Beaufort, Sabine, John Ross, Michael Faraday, John Frederic Daniell, Peter Mark Roget, Charles Wheatstone, and William Snow Harris.[34] The expedition was principally intended as an investigation into terrestrial magnetism, but other sciences were pursued, including study of the tides, the figure of the earth, and meteorology. Meteorology was given greater emphasis than was necessary simply to correct the performance of the magnetic instruments.[35] The committee additionally advised on the instruments with which the naval expedition should be equipped. In terms of meteorology, these included actinometers, Lind's rain gauge, an Osler anemometer, and spirit thermometers for operation in Antarctic temperatures below those at which mercury freezes and mercurial thermometers became ineffective.[36] Procedures were recommended for the verification of the instruments, especially when the expedition was far from fixed observatories on land.[37] Both of the ships – HMS *Erebus* and HMS *Terror* – were to carry standard barometers and thermometers against which others were to be compared. This was especially important when instruments were taken ashore, 'so as to detect and take into account of any change which may have occurred in the interval'.[38] The standards on one ship were to act as checks upon the other.

The passage of the *Erebus* and *Terror* from the tropics to the Antarctic presented an opportunity to investigate von Humboldt's claim that atmospheric pressure at the equator was uniformly 'less in its mean amount than that at and beyond the tropics', a phenomena that was, in turn, believed to produce the trade winds.[39] The observation of changes in the barometer when approaching the line was therefore of great scientific value, as was the observation of the local effects that continents or oceanic currents had on atmospheric pressure. Periods spent at high southern latitudes also presented opportunities to calibrate the instruments. For instance, Ross was asked to verify and to register the ships' standard thermometers at the freezing point of mercury whenever the opportunity arose. This was to be effected by placing four permanent marks on the tube of each standard thermometer, and Ross was 'requested occasionally to compare these

34   Joint Committee of Physics and Meteorology; 1838–39, 19 June 1839, CMB/284, Archives of the Royal Society, London.

35   Royal Society, *Report of the President and Council of the Royal Society on the Instructions to be Prepared for the Scientific Expedition to the Antarctic Regions* (London: Richard and John E. Taylor, 1839), 13.

36   Ward and Dowdeswell, 'On the Meteorological Instruments', 459.

37   Joint Committee of Physics and Meteorology; 1838–39, 22 August 1839, CMB/284, Archives of the Royal Society, London.

38   Royal Society, *Report ... on the Instructions ... for the Scientific Expedition to the Antarctic Regions*, 13.

39   Ibid., 14.

marks with the degrees of the ivory scale'. A bottle of mercury was ordered to accompany each standard thermometer.[40]

The scientific instructions presented to Ross contained, in Ross's words, 'a detailed account of every object of inquiry which the diligence and science of the several committees of that learned body could devise'.[41] This report became a standard for subsequent scientific guides. Ward and Dowdeswell note that the Admiralty's 1849 *Manual of Scientific Enquiry* was effectively a reworking of the 1839 report, prefaced and edited by Herschel.[42]

The deployment of philosophical instruments, and the supply of precise instructions for observations and experiments was not alone enough, however, to guarantee reliable inscriptions. The directions provided to the captains of scientific expeditions were often aspirational in tone and susceptible to compromise when in the field. The robustness of the scientific outcomes of an expedition relied as much on 'immense chains of delegated trust and labour' as they did on detailed instructions, calibrated instruments and well organised skeleton forms.[43] Instruments could not speak for themselves effectively. The determination of their accuracy relied on the person or persons operating them. Identifying and justifying who was to operate which instruments was a crucial matter in voyages of exploration. For John Ross's 1818 voyage to the Arctic, the Royal Society committee suggested to the Admiralty that Sabine was the 'proper person to conduct certain experiments', accompanied by a Sergeant of Artillery to 'take care of instruments'.[44] The Committee also suggested the inclusion of Fisher – 'a Gentleman of considerable mathematical talent' – while Henry Kater reported that the naval officers John Franklin, Frederick Beechey, and William Parry 'had been most assiduous in acquiring a due knowledge of the use of the Instruments to be employed in the Northern Expedition, and that he considers them fully competent to prosecute the required observations and experiments'.[45]

Despite the various controversies surrounding Ross's 1818 Arctic expedition, the Royal Society again recommended Sabine as a member of William Parry's 1819 Arctic voyage:

> It is of the opinion of this Committee that Capt'n Sabine has shown the greatest possible diligence in making the observations which were intrusted [sic] to

---

40   Joint Committee of Physics and Meteorology; 1838–39, 22 August 1839, CMB/284, Archives of the Royal Society, London.

41   James Clark Ross, *A Voyage of Discovery and Research in the Southern and Antarctic Regions, During the Years 1839–43*, 2 volumes (London, 1847), 1: xxvii.

42   Ward and Dowdeswell, 'On the Meteorological Instruments', 455.

43   Schaffer, '"On Seeing Me Write"', 113.

44   Minutes of Committees of the Royal Society Appointed From Time to Time for Particular Purposes, 12 February 1818, CMB1/25, Archives of the Royal Society, London.

45   Minutes of Committees of the Royal Society Appointed From Time to Time for Particular Purposes, 26 March 1818, CMB1/25, Archives of the Royal Society, London.

his care and the greatest judgement and regularity in his method of recording them. And this Committee therefore suggests the propriety of recommending Capt'n Sabine to the Admiralty in the strongest manner, both as deserving every professional encouragement, and as a proper person to be again appointed to take charge of the Observations to be made in a new Expedition.[46]

The reiteration of instrumental and observational competence was crucial. The practices employed and the vagaries of the instruments' fate 'governed the status of the data they produced and the interpretations they suggested'.[47] The reputation of the observer was intrinsically linked to the data and the instruments: 'To question or doubt results or methodology was to question the character and morality of their creator'.[48]

Reforming Meteorology

After 1820, Royal Society committee members were increasingly chosen on the basis of expertise, whether intellectual or professional. This was also true in other respects, such as over the quality and use of the Society's meteorological instruments. The committee formed in 1822 to study this matter incorporated Thomas Young, William Hyde Wollaston and Henry Kater, together with Humphry Davy, Davies Gilbert, the secretaries Brande and Combe, Babbage, Herschel, as well as Luke Howard and John Frederic Daniell, included given their standing in meteorology and related fields.[49] Amongst other recommendations, the Committee ordered the construction of new instruments for the Society, including two barometers from John Newman, of Lisle Street, London; these were, subsequently, the subject of experiments at the Society in December 1822.[50] The observational regime and the siting of the Society's instruments were also reviewed. At a meeting of the Society's Meteorological Committee in 1827, the astronomers and reformers James South and Francis Baily, along with Beaufort and Herschel, complained over recording forms and the quality and situation of its meteorological instruments. They argued that the 'local situation' of its headquarters at Somerset House did not allow for the

46   Minutes of Committees of the Royal Society Appointed From Time to Time for Particular Purposes, 18 March 1819, CMB1/25, Archives of the Royal Society, London.
47   Schaffer, '"On Seeing Me Write"', 112.
48   Waring, 'The Board of Longitude', 66.
49   Anon, Committee for examining into the state of the Meteorological Instruments belonging to the Royal Society, 10 and 12 December 1822, in Minutes of Committees Appointed From Time to Time for Particular Purposes, CMB/1/25, Archives of the Royal Society, London.
50   Anon, Committee for examining into the state of the Meteorological Instruments.

production of 'any series of meteorological observations of material weight and importance in the present state of the science'.[51]

For Jankovic, 'Whether fairly or not, early nineteenth-century commentators ... erupted with criticisms of a general lethargy that supposedly prevailed in the investigation of weather-systems, of the insufficiency *and* profusion of observations, of the public uselessness of the existing stock of facts, and of the imprecision of means for standardizing and using meteorological instruments'.[52] In his *Meteorological Essays*, John Daniell, Professor of Chemistry at Kings College, London, pointed to the Royal Society's meteorological observations as evidence of the poor science undertaken in England. He extended his criticism to the operations of overseas observatories, where, he claimed, there had been insufficient coordination of efforts, such that their 'labour and perseverance lose more than half their value by the want of a well-digested plan of mutual co-operation'.[53] Concerns about the level of training and expertise of meteorological observers similarly preoccupied James Forbes and William Whewell, who argued that science should centre on precision observations and be conducted by trained personnel. Forbes expressed these arguments in his 1832 report on British meteorology to the BAAS meeting in Oxford. For Forbes, meteorological instruments 'have been for the most part treated like toys', while few of the numerous registers 'which monthly, quarterly, and annually are thrown upon the world' could be expected to afford information useful to the development of the science.[54] The situation was, in his view, so bad as to require 'a total revision upon which meteorologists have hitherto very generally proceeded'.[55]

This troubled history of meteorology at the Royal Society is important given discussions over the deployment of meteorological instruments on Admiralty ships. The review of the Royal Society's own instrumental practices was coincident with the Society's advice to captains and scientific officers onboard exploring expeditions. The composition of the Society's committees on these issues was almost identical. It is reasonable to assume, therefore, that the reform of meteorology at the heart of British science was part of attempts to improve the conduct of science at sea. The difficulties experienced at the Royal Society illustrate the difficulties inherent in the pursuit of an exacting instrumental regime. The committees established to advise Parry, Foster and others laid down scientific agendas and observational practices on the assumption that ships were floating

---

51   Anon, Minutes of the Meteorological Committee, 2 August 1827, Meteorological Committee Minutes and Letters 1830–1837, Domestic MSS Volume III, Archives of the Royal Society, London.

52   Jankovic, 'Ideological Crests', 24.

53   John Daniell, *Meteorological Essays* (London: Thomas & George Underwood, 1823), viii.

54   On the use, and misuse, of meteorological instruments, see Jan Golinski, *British Weather and the Climate of Enlightenment* (Chicago: University of Chicago Press, 2007).

55   Forbes, 'Report', quote from pages 196–7. See also Jankovic, 'Ideological Crests', 24.

observatories. At the same time, criticisms of the Society's own meteorological practices illustrated the challenges of meeting such demands when on dry land. When far away from instrument makers and scientific advisers, onboard a moving ship in challenging conditions, operating personnel had no choice but to 'make up and mend ways of recording and transmitting what they reckoned worth noting'.[56]

Francis Beaufort and Instrumental Cultures on Hydrographic Ships

Scientific and exploring expeditions, such as those discussed above, helped establish precedents for the collection of information about terrestrial physics on board ships. The success of these and other voyages in the first half of the nineteenth century encouraged the belief that military vessels might be employed as floating observatories. In his work on French arctic expeditions, Locher notes that the regular maintenance of the systematic naval watch offered real advantages to science, particularly if officers could be compelled to collect data in addition to the other observations they were required to undertake.[57] Naval officers received training in mathematics, navigation and astronomy and would have been comfortable operating relatively sophisticated precision instruments. For observations to be scientifically useful, however, they had to be made regularly, specific instruments had to be employed and full details had to be supplied about their constitution and conditions of use. Particular reduction protocols and computing methods had to be followed. The situation of an instrument and the state of the atmosphere around it had to be given consideration and recorded so that measurements could be reduced to a virtual common environment. The man who spent much of his career persuading the Admiralty to adopt these procedures in the observation of the weather onboard naval ships was Francis Beaufort (1774–1857).

Beaufort began his naval career on an East India Company ship. He served in the Royal Navy in the Napoleonic Wars, reaching the rank of Captain in 1810. In 1829 he was appointed Admiralty Hydrographer of the Navy, a position he held until 1855. Beaufort has been credited with turning the Hydrographic Office into a world-leader in maritime survey.[58] He also played an active role in the administration and advancement of science in Britain. From 1831, he was head of the scientific branch of the Admiralty Board. He spent much of his career promoting the adoption of his wind scale and a system of weather notation onboard hydrographic, naval, merchant and packet ships, and in encouraging the proper use of barometers and thermometers at sea.

As Admiralty Hydrographer, Beaufort was in a position to promote his meteorological agenda amongst officers on the Royal Navy's surveying ships. His influence is evident in the instructions provided in 1831 to Captain Robert Fitzroy

---

56   Schaffer, '"On Seeing Me Write"', 106.
57   Locher, 'The Observatory, The Land-Based Ship and the Crusades', 498.
58   Friendly, *Beaufort of the Admiralty*, 13.

for the latter's hydrographic survey of the South American coastline. Beaufort ordered Fitzroy to keep a 'steadily and accurately kept' meteorological register in which the wind and weather should be observed using Beaufort's own notations. Barometer readings were to be kept and entered into the register, at 9 am and 4 pm. Temperature readings were to be taken at the same times and the extremes of the self-registering thermometer were to be noted daily. The temperature of the sea at the surface was to be taken and compared to that of the air.[59] Beaufort cautioned Fitzroy that no reflected heat should act on the instruments. This was a challenge onboard a ship at night, the more so given the generally warm atmosphere surrounding the vessel.[60]

These duties were extended to other surveying vessels, but Beaufort struggled to ensure that they were met. The log books from various hydrographic surveys during the 1830s demonstrated that what was observed, and when, varied from ship to ship. Beaufort expressed his frustration at the rather haphazard adoption of his meteorological standards in surveying instructions provided to Lieutenant Edward Barnett in 1837, about to take command of HMS *Thunder*. Beaufort justified the use of the barometer and thermometer on the grounds that they would 'provide authentic data collected from all parts of the world', and aid 'future labourers'. He conceded, however, that those hours of entry 'greatly interfere with the employments of such officers as are capable of registering those Instruments with the precision and delicacy which alone can render meteorologic data useful'.[61] Given that meteorological data's future utility was so uncertain, Beaufort suggested that Barnett should do no more than to record the height of the barometer twice a day, along with the extremes of the thermometer.

Frustrated by the slow pace of change, Beaufort and his assistant, Alexander Bridport Becher, mounted a campaign in the late 1830s to persuade the Admiralty to impose a uniform meteorological culture on its ships. Becher was chief Naval Assistant in the Hydrographic Office.[62] In 1838 he wrote a strongly worded memorandum on the subject. As well as urging officers to make use of the log book to record wind and weather as suggested by Beaufort, Becher noted that 'no seaman in command of a ship ever thinks of going to sea without a barometer or sympiesometer'. Becher pointed to the 'great advantage that would arise from the observations of it being recorded in every weather, not to mention during extreme events such as storms and hurricanes, when 'changes in its height during their

---

59   'Admiralty Instructions for the Beagle Voyage' are included in Appendix One of Charles Darwin, *Voyage of the Beagle* (London: Penguin, 1989 [1839]), 396.

60   Ward and Dowdeswell, 'On the Meteorological Instruments', 461–2.

61   Surveying instructions from Beaufort to Lieutenant Edward Barnett of HMS *Thunder*, 9 December 1837, Miscellaneous Files, UKHO, 26 and 27.

62   Nicholas Courtney, *Gale Force 10: The Life and Legacy of Admiral Beaufort* (London: Review, 2002).

progress and times of change should be carefully noted'.[63] Beaufort also urged officers to use the ship's logboard to record the height of the barometer and the thermometer at least once in every watch, suggesting that additional columns be added for this purpose.

The apparent inability of the Admiralty to enforce the meteorological suggestions of its Hydrographer encouraged Beaufort to make use of the Royal Society. He was aided in this by Lieutenant-Colonel William Reid, a Royal Engineer. Reid's interest in meteorology stemmed from his employment as Resident Engineer on Barbados, and centred particularly on Atlantic hurricanes. He published a book on this, *An Attempt to Develop the Law of Storms*, in 1838, which the Admiralty ordered to be supplied to naval officers. In January 1839 Reid wrote to John Herschel, then Chair of the Royal Society's Meteorological Committee, forwarding a letter from Lord Glenelg (Secretary of State for War and the Colonies). The letter announced the intention of the Admiralty to make additions to the log books of naval ships, including columns for observations of the height of the thermometer and barometer.[64]

Courtney claims that it was the Royal Society's intervention that led to the formal issue of marine barometers to the naval fleet in 1843, a matter administered through the Hydrographic Office.[65] This development was reflected in the 1844 edition of the *Admiralty Instructions*, where captains were told to have the barometer 'carefully suspended in some secure and accessible part of the Ship' (and to note its location at the beginning of the log book), and to make observations at 6 am, noon, 6 pm, and midnight.[66] The BAAS was quick to utilise this new development. In 1845, survey ships on the Home Station were ordered to assist the Association, which was interested in observing meteorological phenomena that affected the British Isles during the autumn. Officers were asked to keep registers of barometric observations during October and November using printed directions and blank forms issued especially, and were required to again do so in 1846.[67]

Interest in the value of meteorological instruments at sea spread beyond the survey fleet. In February 1847 Beaufort received a letter from Sir Henry John Leeke, flag-captain of HMS *Queen*, a 110-gun first-rate ship of the line and the last sailing battleship to be completed before the widespread introduction of steam

---

63   Alexander Becher, Proposals for Improving the Meteorological Registers in the Log Books of HM Ships, 14 November 1838, Minute Book 3/HH11, UKHO, 91–3.

64   3 Jan 1839, Letter from Lt Col William Reid, Royal Engineers, to John Herschel, DM/3, Archives of the Royal Society, London.

65   Courtney, *Gale Force 10*, 242.

66   British Admiralty, *Admiralty Instructions for the Government of Her Majesty's Naval Services* (London: Stationary Office, 1844), 173. What was meant exactly by 'secure and accessible' is not explained.

67   Anon, 2 September 1845 and 6 October 1846, Circulars to Surveyors on the Home Station, LB13/HH18 and LB14/HH19, UKHO Archives, Taunton.

power.[68] Leeke wrote to Beaufort to promote the work of his Major of Marines, David M. Adam, whose knowledge of the barometer and attention to changes in the weather had been of 'great use' to him onboard. Adam was 'half a very clever scientific man', claimed Leeke, and he requested additional meteorological instruments to aid Adam.[69] Attached to Leeke's letter was one from Adam himself, addressed to Leeke although presumably targeted at Beaufort, forwarding his readings of the barometer, thermometer, wind direction and force, and weather while HMS *Queen* was at Plymouth Sound in November and December 1846. Adam requested a hygrometer, anemometer, rain gauge, electrometer, and dipping circle, on the grounds that 'If there is one place where accurate knowledge of [the weather], is more useful than another, that place is a Man of War – on ship-board'. For Adam, instruments 'may give the young officers a scientific turn', and that the serious study of meteorology on a flagship could only lead to 'a more accurate knowledge of that science' throughout the fleet.[70] He promised the Admiralty Lords weekly or monthly meteorological reports in return.

Even with these and other developments, Beaufort remained dissatisfied with the scope and quality of meteorological work in the naval and merchant fleets.[71] Writing to the Admiralty in June 1852, Beaufort argued that naval officers were still not 'imbued with a sense of [meteorology's] importance', despite the ease with which a meteorological culture could be engendered:

> The mere record of the Air, & of the Sea at different depths – the force & direction of the winds – the set of the currents – the fluctuations of the Barometer &c. &c. would be easily procured as they would give but little trouble & require but little skill – & after a little time when the officers became familiar with the instruments & warmed to the undertaking, the nicer observations on the state of the atmosphere, the quantity of rain, the indications of changes in the wind & weather, and minute descriptions of mutual phenomena would be duly registered according to the forms which might be prescribed in what might be called a supplementary log.[72]

---

68   J.J. Colledge, *Ships of the Royal Navy: The Complete Record of all Fighting Ships of the Royal Navy from the Fifteenth Century to the Present* (London: Greenhill Books, 1970).

69   Letter from Sir Henry Leeke to Francis Beaufort, 9 February 1847, Incoming Letters L301-669, UKHO Archives, Taunton.

70   Letter from David Adam to Henry Leeke, 30 January 1847, appended to the letter from Leeke to Beaufort.

71   In 1848 sea trials were carried out at Woolwich of a new wind gauge, designed by Colonel W.H. Rochfort, which he claimed eliminated the effects of the ship's movement on measurements of wind force: Memoranda written by Beaufort from 27 January 1848 to 29 April 1848, MB6/HH21, UKHO Archives, Taunton.

72   Francis Beaufort, Meteorological Observations – General System of Observing By Foreign and Home Men of War – Packets – & Merchant Vessels, 25 June 1852, MB8/HH28, Letterbooks, UKHO, Taunton, 79–80.

Beaufort again turned to the Royal Society, hoping that it could provide a skeleton form, a list of instruments and duplicates to go to each vessel, advice on who should build them, and 'by whom compared & verified'. He also hoped that the Admiralty and the Board of Trade might persuade the directors of mail packet companies and merchant ships respectively to entice their crews to do something similar.

The Brussels and London Maritime Conferences

In 1853 a maritime conference was convened to establish a uniform international system of meteorological observation at sea, its justification being the improvements that could be made to navigation and scientific knowledge.[73] Although the idea for the conference came from William Reid and his former commanding officer, Sir John Fox Burgoyne, it was organised by the US Government and held in Belgium, chaired by the astronomer Adolphe Quetelet. The British delegates were Captain Henry James of the Ordnance Survey and Captain Frederick Beechey, head of the Marine Department of the Board of Trade. The principal aim of the conference was to design a meteorological register for use on naval and, if possible, merchant ships. Delegates were wary of extending the conference remit, although discussion strayed inevitably onto instrumentation. Beechey argued that it was impossible to recommend the adoption of any particular instruments, let alone any specific instrument makers, given that different scales and standards were in use internationally; and that any standardization of instruments would 'interfere too abruptly with long established usages and long established records, with which the observations now to be collected would require a reduction, before they could be compared'.[74]

Despite a general reluctance to standardise instruments and instrumental cultures, there was some debate over the use of the thermometer and barometer at sea. Delegates recommended that ships should carry both, along with 'at least one good chronometer, one good sextant, [and] two good compasses'. The conference acknowledged the widespread use of barometers onboard seagoing vessels of all types and their value as indicators of changes in relative pressure, but their use as recorders of absolute pressure was lamented: 'That an instrument so rude and so abundant in error, as is the marine barometer generally in use, should in this

73   Malcolm Walker, *History of the Meteorological Office* (Cambridge: Cambridge University Press, 2012).

74   Anon, *Maritime Conference held at Brussels for Devising an Uniform System of Meteorological Observations at Sea*, MS, 1853, Archive of the Royal Meteorological Society, Meteorological Office, Devon, 60.

age of invention and improvement be found on board any ship, will doubtless be regarded hereafter with surprise'.[75]

Similar sentiments were expressed of the thermometer. The conference noted that too many were in use at sea despite users having no idea of their degree of error, which rendered the results worthless. The French delegate, Alexandre Delamarche, argued that a uniform thermometer should be adopted internationally, but, as for the barometer, he acknowledged the difficulty of introducing a single universal design. It was, nevertheless, agreed that the centigrade scale should be added to all thermometers (although not to the thermometer attached to the barometer), alongside any other scale currently in use. This was justified on the grounds of the possible *future* adoption of the centigrade scale – 'to accustom observers in all services to its use' – rather than its immediate use; the conference rejected the proposal that a separate centigrade column should be added to the meteorological register.[76]

The immediate consequences of the conference were positive, at least as far as the British contingent was concerned. Beechey reported on the outcomes of the conference to the British government and, in February 1854, the First Lord of the Admiralty, Sir James Graham, announced in the House of Commons that a new government department was to be formed, called the Meteorological Department, to be funded through the Board of Trade and the Admiralty. Its aims were to effect the recommendations of the conference: to collect and analyse meteorological observations taken at sea; to promote the observation of the weather on board ships; and, in the spirit of international cooperation, to convey reduced observations to the US Naval Observatory.[77] This Department, led by Robert Fitzroy, began to supply instruments, instructions and registers to Royal Navy ships and British merchantmen, and to collect and compile weather logs. The Department was not long in attracting criticism, notably from Francis Galton, the African explorer, meteorologist and eugenicist, who took issue with FitzRoy's attempts at weather forecasting.[78] The workload and criticism took its toll on FitzRoy, who committed suicide on Sunday 30 April 1865. The committee of inquiry formed to consider the Department's work in the aftermath of FitzRoy's death was chaired by Galton and identified several shortfalls and proposed some suggestions for improvement. While Galton's Report complimented the Department on overseeing the provision

---

75   Anon, *Maritime Conference*, 18. In relation to relative pressure, the conference noted the value of the aneroid barometer at sea but preferred the more delicate mercurial barometer given its ability to provide absolute results.

76   Anon, *Maritime Conference*, 14.

77   Walker, *History of the Meteorological Office*, 21–2.

78   Nicholas Wright Gillham, *A Life of St Francis Galton: From African Exploration to the Birth of Eugenics* (Oxford: Oxford University Press, 2001).

of ships with instruments and registers, it was felt that too few registers had been collected and that there was insufficient global coverage.[79]

There were several attempts to develop trans-national networks of meteorological observations and stations in the 1860s, but these proved unsuccessful, due in part to the unstable political situation in Europe.[80] Several conferences on the topic were eventually convened in the 1870s. The first was held in Leipzig in 1872, followed by an international congress in Vienna in 1873. In 1874 a private conference on maritime meteorology was held in London. The London conference set out to review participating nations' implementation of the recommendations of the 1853 Brussels conference, and to promote the recommendations of the 1872 and 1873 meetings; namely, that 'Thorough uniformity in methods and instruments should be aimed at'; 'Unity of measures and scales is desirable, and to this end the introduction of millimètres for the barometer and the Centigrade scale for the thermometer should be aimed at'; and that 'the importance of the co-operation of the Navies' should be promoted.[81] In its aims the conference was part of wider movements in the 1870s to establish international standards, such as the international gold standard and the Treaty of the Metre.[82]

Participants' responses revealed differences of opinion on the aims and successes of the 1853 conference and over subsequent attempts to introduce a uniform international approach to the study of meteorology at sea. Brigadier-General Myer, Chief Signal Officer in the US Army, reported that the USA had followed the Brussels plan. J.C. de Brito-Capello, Director of the Nautical and Meteorological Observations at the Lisbon Observatory, noted the same of Portugal. The Danish were also supportive, although Captain Hoffmeyer of the Danish Royal Meteorological Institute conceded that some compromises had been made, such as the use of aneroid barometers on smaller vessels where a mercurial barometer 'cannot appropriately be placed'.

Other nations were less positive. Professor Buys Ballot, the Dutch meteorologist and the meeting's President, argued that the Brussels conference had 'asked for too many observations' and that the hours of observation were inconvenient. The French made similar complaints. The report of Captain Rikatcheff of the Imperial Russian Navy was perhaps the most pessimistic. The thermometers used on board Russian vessels had continued to use the Reaumur scale, without the recommended

---

79   J. Burton, 'Robert FitzRoy and the Early History of the Meteorological Office', *British Journal for the History of Science* 19 (1986): 147–76.

80   Walker, *History of the Meteorological Office*, 109.

81   Meteorological Committee, *Report of the Proceedings of the Conference on Maritime Meteorology held in London, 1874* (London: HM Stationary Office, 1875), 4.

82   Martin H. Geyer, 'One Language for the World: The Metric System, International Coinage, Gold Standard, and the Rise of Internationalism, 1850–1900', in Martin H. Geyer and Johannes Paulmann (eds) *The Mechanics of Internationalism: Culture, Society, and Politics from the 1840s to the First World* War (Oxford: Oxford University Press, 2001), 55–92.

addition of the centigrade scale, because of worries that 'one would often be read instead of the other'. The barometers had not been compared since 1853 and the necessary corrections not been determined, due to the want of a dedicated office to do so. Rikatcheff complained that meteorological observations obtained at sea were not discussed or utilized in his country, or shared with other countries. For Rikatcheff, 'You ought not to be astonished, Sir, if from these answers you see that the greater part of our Maritime Meteorological Observations lie dormant till now'.[83]

Discussion of the various recommendations made at the 1874 conference was similarly wide-ranging and conflicting. Disagreements remained over which scales to use when measuring temperature; to what degree of accuracy readings should be taken; what scale should be used to record wind force; how the labour of global meteorological study should be divided; what form the meteorological register should take; and how the resultant data should be dealt with, analysed and archived. The various formal conference resolutions reflected these differences.[84] Attempts to fashion a unifying, international language of science, which would allow meteorologists of all nations to communicate with one another, was part of a wider effort to effect a system of liberal internationalism in the 1870s; to foster economic and social progress; and popularize a language of progress. As with concurrent attempts to encourage the universal adoption of the metric system, however, the implementation of a single international system of marine meteorology was stymied by national rivalries and resistance to new measures and practices on board the ships of the various European navies.

Conclusion

This chapter has shown that during the nineteenth century Britain's Admiralty gradually adopted and honed the practice of observing and recording the weather on board its ships. The polar expeditions of the 1810s and 1820s served as important test sites for the development of new observation protocols, registers and instrumental practices, which, in turn, were informed by scientific reforms then taking place. These practices were then introduced more widely, initially to the Hydrographic Office's survey ships and then to the naval fleet.

The implementation of a universal schedule of observations and the correct siting and use of various instruments onboard surveying and fighting ships was not easily achieved. Beaufort and others struggled to persuade the Admiralty of the value of systematic meteorological observations on its ships. Getting officers to conform

---

83   Meteorological Committee, *Report of the Proceedings. Appendix B*, quotes from pages 28, 26, 31, and 32 respectively.

84   In an attempt to encourage the more uniform pursuit of maritime meteorology, the 'Proposed English Instructions for keeping the Meteorological Log' were appended to the conference proceedings.

to protocols once they were laid down remained a challenge, especially on ships whose main purpose was not scientific or hydrographic. Observations were made at irregular times of the day and instruments were sited inappropriately. Observers failed to check instruments against standards, entered readings into the registers retrospectively and made erroneous reductions. Attempts in the later nineteenth century to institute a single meteorological culture across European navies further exposed the highly localized nature of a supposedly universal science.

It is not, then, straightforwardly possible to conceive of British naval and surveying vessels as itinerant observatories, if, that is, we assume the observatory to have been an uncompromised site for science and its observers obedient drudges. The evidence examined here makes it difficult to agree entirely with Edwards' claim that mid-nineteenth-century meteorology conformed to or was productive of a quasi-obligatory globalism. Scientific manuals and conference proceedings give the impression of international consensus, but closer reading reveals differences over issues of instrument type, the scales to use, when and how to observe, and so on. All that said, if we assume that observatories and their instruments always existed in various states of repair and that observational practices internal to them were always liable to compromise and dissent, and if we proceed on the basis that international scientific networks were always performative and subject to translation, then it does become possible to include the man-of-war and its various barometers, thermometers and log books as critical components in the extension of European power over space.[85]

---

85   I am grateful to the British Academy for the research funding on which this chapter is based.

Chapter 5

Objects of Exploration: Expanding the Horizons of Maritime History

Claire Warrior and John McAleer

On the afternoon of Monday 1 May 1769, during HM Bark *Endeavour*'s voyage to the Pacific Ocean to observe the Transit of Venus, James Cook recorded that the shore party 'set up the Observatory and took the Astronomical Quadrant a shore [sic] for the first time, together with some other instruments'. The following morning, around 9 o'clock, Cook was aghast to discover that the quadrant was 'not to be found'.[1] Cook's account of the disappearance of the instrument, which was purloined by an indigenous inhabitant of Tahiti while 'most of our people ... were diverting themselves with the natives', has been interpreted as a key moment in the early history of encounters between Europeans and Pacific islanders.[2] In this context, the quadrant, which 'had never been taken out of the packing case (which was about 18 inches square) since it came from Mr Bird the Maker', acts as a proxy, symbolising broader kinds of cultural misunderstandings that characterise the nature of European exploration in the eighteenth-century Pacific Ocean.[3]

At the heart of the incident is a story about how scientific investigation was conducted on voyages of exploration. In recounting the disappearance of the quadrant, Cook provides details about the context in which he and his companions intended to use it. Perhaps the most remarkable thing in the account of the stolen quadrant is the highly fortified observatory that the British travellers constructed to house their apparatus: 'The north and south parts consisted of a bank of earth 4½ feet high on the inside, and a ditch without, 10 feet broad and 6 feet deep: on the west side facing the bay a bank of earth 4 feet high and palisades upon that ... on the east side upon the bank of the river was placed a double row of casks: and as this was the weakest side the 2 four pounders were placed there'.[4]

---

1   James Cook, *The Journals of Captain James Cook on his Voyages of Discovery. Volume I: The Voyage of the 'Endeavour', 1768–1771* (ed.) J.C. Beaglehole (Cambridge: Cambridge University Press, 1955), 86, 87.

2   Ibid., p. 88. For further discussion of this interpretation, see Richard C. Allen, '"Remember me to my good friend Captain Walker": James Cook and the North Yorkshire Quakers', in *Captain Cook: Explorations and Reassessments* (ed.) Glyndwr Williams (Woodbridge: Boydell, 2004): 21–36, quote from page 34.

3   Cook, *The Voyage of the 'Endeavour'*, 87.

4   Ibid., 86–7.

How far is this fortification, or its component parts, also an instrument of exploration? Can we regard the men who operated the precision instruments, or who defended the mobile observatories, as instruments of exploration too? This chapter seeks to illuminate some of these questions and to offer a more eclectic, if less precise, definition of what it means to be an instrument of scientific investigation in the age of European exploration. Do 'traditional' scientific instruments, like Bird's quadrant, comprise all of the instruments of exploration at Cook's disposal? Can objects that fall outside this category be considered tools of scientific exploration? Are there other ways of defining the tools of exploration that helped Europeans to chart, map, survey, and know places like the Pacific and the Poles? Are the paintings produced on voyages of exploration, or the objects acquired during them, or the publications derived from them also 'instruments' that might be used to collect and disseminate scientific data? Can personal effects, ethnographic items, or even entire vessels be categorised and interpreted as instruments of exploration?

In attempting to interrogate what we understand by the term 'instruments of exploration', and to explore its elasticity, this essay draws on the work of a range of scholars across several disciplines. It acknowledges the recent work by a number of geographers, historians, and anthropologists that has extended our concept of exploration to encompass, in the words of Daniel Clayton, 'a complex cultural and spatial practice that mobilised a range of material and imaginative resources (equipment, patronage, publicity, scholarship, myths) and that stretched well beyond the initiatives and texts of individuals'.[5] The essay is also indebted to the increasing scholarly awareness of, and interest in, the interactions between science and material culture.[6] And finally, it acknowledges well-established anthropological ideas about the cultural biographies of things, in which the identity of an artefact is understood to be situational.[7] Cook's quadrant, for example, moved rapidly between being a scientific instrument and a stolen treasure during the expedition, and is now a museum artefact.[8]

The chapter uses the collections at the National Maritime Museum, Greenwich (NMM), and their display and interpretation, as a case study. In addition to

---

5   Daniel Clayton, 'Captain Cook's Command of Knowledge and Space: Chronicles from Nootka Sound', in *Captain Cook* (ed.) Williams, 110–33, quote from page 111.

6   Sujit Sivasundaram, *Nature and the Godly Empire: Science and Evangelical Mission in the Pacific, 1795–1850* (Cambridge: Cambridge University Press, 2005), 9, n. 31.

7   Igor Kopytoff, 'The Cultural Biography of Things: Commoditization as Process', in *The Social Life of Things: Commodities in Cultural Perspective* (ed.) Arjun Appadurai (Cambridge: Cambridge University Press, 1986), 64–91. For an application to a particular artefact, see Laura Peers, '"Many Tender Ties": The Shifting Contexts and Meanings of the S BLACK Bag', *World Archaeology* 31 (1999): 288–302.

8   The Royal Society claims that a Bird quadrant amongst its collections was used by Cook during his observations of the transit of Venus in Tahiti: see https://pictures. royalsociety.org/image-rs-8472, whilst the Science Museum also has at least one quadrant (object 1876–572) that is said to have been used on the same voyage.

providing a focus for our discussion, it replicates the sorts of contexts in which the general public can engage with and understand historical 'instruments of exploration'. In other words, the public encounters such scientific instruments in conjunction with larger interpretative frameworks that try to convey information about the history of science more generally, its connections with the history of British maritime power, narratives of national identity, naval hagiography, and so on. In addition to contributing to academic debates, we hope to suggest ways in which more latitude in the definition of terms such as 'instrument of exploration' might offer a means of introducing the history of science to larger audiences, as well as opportunities to use the history of science and historical geography to inform the disciplines of maritime history, imperial history, the history of collecting, and art history. Focusing on the materiality of exploration enables us to explore the correspondence between what happened in the field, what was brought home, and how such things have been understood. Instruments, after all, have 'lives' in many locations and contexts. Focusing on the materiality of exploration provides another perspective on the complex ways in which the actual embodied experience of exploration has informed, or existed in tension with, the scientific results of such expeditions. We hope that this focus on the interplay between subjects and objects will inspire further research on key issues relating to exploration: what is it? Where does it take place? What and who is involved in conducting it?

The chapter begins by providing a brief overview of the history of the NMM, offering some initial thoughts about the collections as a whole. It then addresses specific types of 'instruments' that fall outside the traditional categories of precision technology but which, nevertheless, might be used to shed light on the ways in which British travellers and explorers engaged with the wider world from the eighteenth century to the twentieth. To the predictable list of chronometers, telescopes and sextants we suggest adding categories such as ships and art. In conclusion, the chapter explores how the cultural practices of exploration and explorers themselves might be considered as part of this wider definition of 'objects of exploration'. Recent reconceptualisation of exploration as a practice and a culture might, we suggest, also extend to the role of instruments.

The National Maritime Museum

The National Maritime Museum was founded by Act of Parliament in 1934 and opened to the public in 1937. The museum's name was chosen by Rudyard Kipling, one of its early champions. Kipling hoped to encapsulate and commemorate the wide variety of people and events who had contributed to Britain's history as a maritime nation: 'the work of the great explorers, the chart-makers, the adventures of ocean pioneers, the traffic of sea-traders, the discoveries of those who devised safer methods of traversing the ocean-ways, the romance of the great liners, the forgotten struggles of the tramp-steamers, and ... the domestic work of the fishers plying their nets in peace and war'.[9] His vision was wide-ranging and suitably ambitious for an institution dedicated to conveying this key aspect of British history. Among its many other remits, the NMM has exhibited the history of empire and exploration. In the absence of a 'National Museum of British History', the NMM has, in many ways, filled this void. This is perhaps unsurprising. The Museum is located on the site of the former Royal Hospital School, once known as the 'cradle of the Navy'. The teenage home and burial place of General James Wolfe is nearby. The Museum incorporates collections relating to Sir Francis Drake, Captain James Cook, Admiral Lord Nelson, and Sir John Franklin among others. In displaying the history of the Royal Navy and merchant fleets, voyages of exploration, the race to the Poles, and the story of London as a port, the NMM encompasses many of the essential elements of Britain's imperial history. And it was founded at a moment when its supporters believed that the establishment of such an institution, and the consequent instilling of 'legitimate pride' in its visitors, might help to stem the decline in Britain's political fortunes in the period.[10] There was also something of a salvage mind-set to this endeavour: the perception of the time was of a maritime heritage rapidly disappearing from the national grasp.[11] From its earliest days then, the NMM has collected, displayed and interpreted Britain's intertwined national and imperial history.[12]

The addition of the Royal Observatory to the NMM in 1957 further strengthened the link between Britain's maritime history and the country's impact on the wider world through scientific and geographical exploration. In addition to distributing time to the world in the form of Greenwich Mean Time, the Royal Observatory at Greenwich also checked and calibrated the precision instruments that Royal Navy explorers took to the furthest ends of the globe on their voyages. The history of science is, therefore, inextricably connected with histories of exploration, empire and British maritime endeavour. This is even more evident in the collections that comprise the museum.

The NMM has the largest collection of objects of any maritime museum in the world.[13] It holds over two and half million items, ranging across all sorts of materials and media. These include oil paintings, ship models, textiles, jewellery, coins and medals, weapons, and prints and drawings, as well as an unrivalled

---

9   Quoted in Kevin Littlewood and Beverley Butler, *Of Ships and Stars: Maritime Heritage and the Founding of the National Maritime Museum, Greenwich* (London: The Athlone Press, 1998), 52.

10   Ibid., xv.

11   See, for example, the comments of naval historian John Leyland, who described the export of artworks to Germany and the USA, along with the loss of national relics into 'careless hands'; quoted in Littlewood and Butler, *Of Ships and Stars*, 25.

12   See also Phyllis Leffler, 'Peopling the Portholes: National Identity and Maritime Museums in the U.S. and U.K.', *The Public Historian* 26 (2004): 23–48.

13   'Collection Development Policy, 2012–17', available at http://www.rmg.co.uk/upload/pdf/Collection_Development_Policy_2012-2017.pdf, 10.

collection of manuscripts covering practically every aspect of British exploration. The museum is also fortunate in holding many iconic objects relating to famous people, including explorers such as James Cook, John Franklin, and Robert Falcon Scott, many of whom have played their own heroic role in the national psyche. The scientific instruments collection, as the chapter by Richard Dunn in this volume makes clear, is a crucial part of the museum's holdings. However, is it possible to extend the idea of an 'instrument'? Current curatorial practice seeks to reconnect objects with the practices that they facilitated, the wider contexts in which they were used and the people that manipulated them. Rather than focus solely on the technical details or physical qualities of things, museum professionals now strive to move beyond the artefact and to think about the experience of those using them. In the context of scientific instruments, these developments have prompted curators to consider how and where navigation, surveying, charting and knowledge-gathering took place. Were such practices carried out solely using scientific instruments as traditionally defined? Did other objects (or types of object) contribute to such activities and, if so, how? Perhaps the first place to look in attempting to answer this is the vehicle that carried these precision instruments and made exploration possible: the ship.

The Ship

Technically, the ship is not one of the NMM's official collection 'categories', other than in the miniaturised form of ship models. And yet it is the classic maritime 'object', and one intimately connected with the practice of exploration.[14] Ships enable forms of control and the projection of power across vast distances.[15] They connect people and places. They bring together a range of sophisticated technologies into one unit: ships were the vehicles used to transport large quantities of scientific objects, personnel and the items that sustained them around the world on voyages of European exploration.[16] In short, ships can be seen, as Richard Sorrenson points out, as scientific instruments.[17] They were not just places where instruments were stored but places in which knowledge and expertise coalesced. If, as some scholars have recently suggested, we might regard ships as self-contained maritime communities, then perhaps ships on scientific voyages of exploration might be seen as floating universities or academies where resources and expertise

---

14   John Mack, *The Sea: A Cultural History* (London: Reaktion, 2013), 136–65.

15   For more on this, see Carlo Cipolla's classic, *Guns and Sails in the Early Phase of European Expansion, 1400–1700* (London: Collins, 1965).

16   For further discussion, see Clayton, 'Captain Cook's Command of Knowledge and Space', 110–33.

17   Richard Sorrenson, 'The Ship as a Scientific Instrument in the Eighteenth Century', *Osiris* 11 (1996): 221–36.

were gathered together and experimentation was carried out.[18] In their time, these ships acted as the world's first effective mobile laboratories, where newly acquired data and previously accumulated and computed knowledge were related to each other in the actual process of discovery.[19]

The importance of the ship to the entire enterprise of exploration was not lost on those who named them. Vessels often carried connotations of investigation, exploration and endeavour (literally in the case of Cook's first voyage). For example, in December 1771, on the eve of the departure of Cook's second voyage, his ships were christened *Adventure* and *Resolution* respectively, names that Cook felt were more appropriate to the task being undertaken than the original names chosen by the Admiralty.[20] This was even more obvious in the case of the subsequent voyage of one of Cook's lieutenants, Matthew Flinders. At the end of November 1800, the 30-foot long sloop, *Xenophon*, was selected and renamed *Investigator*. With a complement of 80 men under Lieutenant Flinders, it sailed from Spithead in July 1801 and arrived in Australia in December to begin its circumnavigation of the continent.[21]

Ships' names were not the only indicator of their status as tools aiding exploration: there were more concrete contributions. The example of Flinders's *Investigator* is again instructive in this regard. It was one of the most thoroughly equipped survey vessels to leave Britain, carrying sextants by Jesse Ramsden, chronometers by Arnold and Earnshaw, and a host of other surveying and astronomical instruments by Edward Troughton. There was even a greenhouse for live plants on the quarterdeck. The ship acted as a receptacle for these objects and it brought them together, making them potentially greater than the sum of their parts. Similarly, the people on board offered a range of skills and expertise. In the *Investigator*, these included Robert Brown (botanist), John Crosley (astronomer), William Westall (landscape painter), Ferdinand Bauer (draughtsman), Peter Good (gardener), and John Allen (miner).[22] Their creative interaction on board such an

---

18   On ships as floating maritime communities see, for example, Marcus Rediker, *Between the Devil and the Deep Blue Sea: Merchant Seamen, Pirates and the Anglo-American Maritime World, 1700–1750* (Cambridge: Cambridge University Press, 1989); Marcus Rediker, *The Slave Ship: A Human History* (London: John Murray, 2008); Emma Christopher, *Slave Ships and their Captive Cargoes, 1730–1807* (Cambridge: Cambridge University Press, 2006).

19   See Bernard Smith, *Imagining the Pacific: In the Wake of the Cook Voyages* (Carlton, Victoria: Melbourne University Press, 1992), 49.

20   Nicholas Thomas, *Discoveries: The Voyages of Captain Cook* (London: Penguin, 2004), 144. The Admiralty worried that the names *Raleigh* and *Drake* might be offensive to the Spanish.

21   For further details, see Juliet Wege, Alex George, Jan Gathe, Kris Lemson, and Kath Napier (eds) *Matthew Flinders and his Scientific Gentlemen* (Welshpool, WA: Western Australian Museum, 2005).

22   David Mackay, *In the Wake of Cook: Exploration, Science and Empire, 1780–1801* (London: Croom Helm, 1985), 4.

academy of science was a key part of the scientific process. In this, the *Investigator* mission was modelled on and closely mirrored the situation on Cook's various voyages, where a range of European intellectuals, such as Anders Sparrman and the Forsters, shared their living space with men of different professional backgrounds and intellectual training. The ship as a nexus for collating knowledge continued in the nineteenth century. HMS *Rattlesnake*, which left Portsmouth on 30 November 1846 on a voyage to northern Australia and New Guinea, was similarly equipped. It carried 39 cases of instruments, including 28 chronometers, compasses and a number of devices for magnetical observations. It also travelled with expert passengers such as Thomas Henry Huxley (marine naturalist), John MacGillivray (botanist), and Oswald Brierly (artist).[23]

These vessels acted as repositories for 'traditional' scientific precision instruments, bringing them together as a critical mass. But, among the other objects carried by these ships were things that might not be immediately obvious as instruments of scientific exploration. The most mundane of objects – such as paper and books – were vital to the recording work of artists and the computation tasks and study of travelling men of science. When Joseph Dalton Hooker left Britain on the *Erebus* in 1840 under James Clark Ross, he recorded the kind of material and equipment that he needed for his work: 'As botanist my outfit from Government consisted of about 25 reams of paper, of three kinds – blotting, cartridge, and brown; also two botanising vascular and two of Mr Ward's invaluable cases for bringing home plants alive, through latitudes of different temperatures. I was further, through the kindness of my friends, equipped with botanical books, microscopes etc. to the value of about £50, besides a few volumes of Natural History and general literature'.[24]

In addition to precision instruments, the principal requirement for the kind of scientific work being conducted on voyages of exploration was perhaps the reference library. On the first Cook voyage, for instance, an account of the circumnavigation of Admiral George Anson, published in 1748, was taken on board the *Endeavour*, as both Cook and Joseph Banks refer to it.[25] This was not an isolated incident. Banks carried a further 60 titles, including accounts of the principal voyages to the Pacific prior to Cook, major works on natural history, and the travels of natural historians to regions beyond Europe such as Brazil (Pies and Marcgraf) and Jamaica (Sloane). These texts provided him with a model for conducting his research in the Pacific.[26] In addition to the impressive array of precision technology and expertise noted above, the voyage of the *Investigator* also incorporated other forms of scientific instrument. The expedition was under

---

23   Jordan Goodman, *The Rattlesnake: A Voyage of Discovery to the Coral Sea* (London: Faber and Faber, 2006), 35.

24   Quoted in Jim Endersby, *Imperial Nature: Joseph Hooker and the Practices of Victorian Science* (Chicago: University of Chicago Press, 2008), 59.

25   Smith, *Imagining the Pacific*, 53.

26   Ibid., 43.

the patronage of Banks – then President of the Royal Society – and he ensured that it was well equipped with a comprehensive library of charts and relevant literature. Banks lent his copy of the *Endeavour* log from Cook's first voyage. Prior to sailing, Flinders requested a range of charts including Laurie and Whittle's *Complete East India Pilot* and Arrowsmith's charts of the world and of the Pacific. He also sought material on the voyages of Dampier, Dalrymple, Cook, Hawksworth, and Bligh, and asked for Murdoch McKenzie's *Treatise on Maritime Surveying* (1797). He additionally requested a set of *Encyclopaedia Britannica*, presumably the third edition in 15 volumes published between 1788 and 1797. As these examples demonstrate, to go on a scientific voyage in a naval ship was to sail with a wide range of relevant, up-to-date knowledge and equipment ready to hand. In housing a plethora of instruments and exploration aids, ships actively facilitated the gathering of knowledge about the world's oceans and coastal littorals. As floating laboratories, where data were collated and results discussed, ships played a central role in the practices of exploration, the acquisition of knowledge and its safe return to dry land.

Stepping ashore briefly, it is important to note that published accounts and written records were also vital to the work of exploration after the return of a voyage.[27] They became instruments for laying claim to exploration achieved and research conducted. Masters and captains kept official logs. But in the course of the eighteenth century, the logs of scientific travellers were also increasingly prized as part of the expedition's findings. At the end of his first voyage, for example, Cook urged that its results should be 'published by authority to fix the prior right of discovery beyond dispute'.[28] Archibald Menzies, the botanist on George Vancouver's mission to the north Pacific in the 1790s, was instructed to keep a journal to be submitted to the Secretary of State at the end of the voyage.[29] Conversely, the absence of a published account could do untold damage to reputations and the perceived veracity of exploration voyages. The importance of publication as a guarantee of accuracy is borne out by the controversy surrounding Vitus Bering's voyage of 1741 (the Second Kamchatka or Great Northern Expedition). Doubts about its significance surfaced partially because the Russian authorities failed to publish any proper account of the voyage until 1758.[30] Similarly, Alessandro Malaspina's achievements in the same region of the

27   On this issue, see Charles W.J. Withers and Innes M. Keighren, 'Travels into Print: Authoring, Editing and Narratives of Travel and Exploration, *c.*1815–*c.*1857', *Transactions of the Institute of British Geographers* 36 (2011): 560–73, and Innes M. Keighren, Charles W.J. Withers and Bill Bell, *Travels into Print: Exploration, Writing, and Publishing with John Murray, 1773–1859* (Chicago: University of Chicago Press, 2015).

28   Glyn Williams, *Voyages of Delusion: The Search for the Northwest Passage in the Age of Reason* (London: HarperCollins, 2003), 354.

29   Mackay, *In the Wake of Cook*, 102–3.

30   Williams, *Voyages of Delusion*, 247.

north Pacific are still relatively unknown because, for a host of complex political reasons, his account lay unpublished for almost 100 years.[31]

Works of Art

One of the early ambitions of the founders of the National Maritime Museum was to use paintings to tell the visual story of Britain's maritime history. This focus on the visual might partially explain why the museum has some 4,000 oil paintings and 60,000 prints in its collection, and why its earliest displays centred on works of art rather than artefacts.[32] Among the jewels of the collection are the paintings of the great voyage artists that accompanied James Cook and Matthew Flinders: Sydney Parkinson, William Hodges, John Webber, and William Westall. These works of art frequently found their way on to the walls of the Royal Academy. Prints and drawings derived from them were aimed for publication and wider dissemination in the narratives of exploration that proliferated in the period. Today, their work is often used by publishers, publicists, and even museums to illustrate exploration in progress and science in action, to show scientific instruments being used, and scientific activities being undertaken.

But is this the limit of their interpretation? Are they only useful as outliers in the canon of eighteenth-century British art history, or as illustrations in some grand narrative of British maritime history? They can also be regarded, we suggest, as instruments of exploration in their own right. The instructions to Cook, outlining the value of a visual record, makes this clear. The Lords of the Admiralty were adamant that art should play a practical and functional role in the mission. Those artists accompanying Cook were required 'to make drawings and paintings of such places in the Countries you may touch at in the Course of the said Voyage as may be proper to give a more perfect idea thereof than can be formed from written descriptions only'.[33] The first requirement of the voyage was that it should return with accurate records, and the art of people like William Hodges (and others) played a key role in this process of scientific record-making. Cook himself reminds us that, in some of the situations in which he found himself, 'the whole exhibits a view which can only be described by the pencle [sic] of an able

---

31   Robin Inglis, 'Successors and Rivals to Cook: The French and the Spaniards', in *Captain Cook* (ed.) Williams, 161–78, quote from page 177. See also John Kendrick, *Alejandro Malaspina: Portrait of a Visionary* (Montreal: McGill-Queen's Press, 2003).

32   For further details, see Littlewood and Butler, *Of Ships and Stars*; Geoff Quilley, *Empire to Nation: Art, History and the Visualization of Maritime Britain, 1768–1829* (New Haven and London: Yale University Press, 2011).

33   Bernard Smith, 'William Hodges and English *Plein-air* Painting', *Art History* 6 (1983): 143–52, quote from page 145.

painter'.[34] During the 12 years from 1768 to 1780, around 3,000 original drawings were made of things mostly from the Pacific, not seen before by Europeans: plants, fish, molluscs, birds, coastlines, landscapes, unknown peoples, their arts and crafts, religious practices and style of life.[35] These visual records were contributions, albeit partial and incomplete, to the imperial archive of information being steadily accumulated throughout the eighteenth century.[36]

William Hodges's work provides an interesting case study in this regard. Hodges accompanied Cook on the latter's second voyage to the Pacific, between 1772 and 1775. His coastal profiles, and his capturing of 'typical' climatic phenomena demonstrate how fine art could act as an instrument of exploration. At the outset, his classical fine art training gave little indication of his ability to turn his hand to scientific recording of landscape phenomena.[37] Yet, in Hodges's work on Cook's voyages, works of art were more than just aesthetically pleasing assemblages of lines and tones: these visual records transcribed, fixed and recorded landscapes and seascapes for posterity, providing a rich visual archive of data for future travellers and anyone else who might be interested. They acted, in short, as instruments of exploration.[38] Hodges's work has been described by Harriet Guest as comprising 'images in which the elaborately theorised conventions of high art mesh with the conflicting demands of natural history for the accurate representation of climate, topography, flora and fauna'.[39] The representation of atmospheric effects indicates how aesthetically conceived landscape views could be simultaneously permeated by the discourse of scientific enquiry.[40]

Hodges's *View of Table Mountain from Table Bay* provides a tangible example of this approach (Figure 5.1). It combines the naval tradition of the topographically

---

34   Diary entry, Wednesday, 24 February 1773, in James Cook, *The Journals of Captain James Cook on his Voyages of Discovery. Volume 2: The Voyage of the 'Resolution' and 'Adventure', 1772–1775* (ed.) J.C. Beaglehole (Cambridge: Cambridge University Press, 1961), 98–9.

35   Smith, *Imagining the Pacific*, 51.

36   Thomas Richards, *The Imperial Archive: Knowledge and the Fantasy of Empire* (London: Verso, 1993).

37   John Bonehill and Geoff Quilley (eds) *William Hodges, 1744–1797: The Art of Exploration* (New Haven: Yale University Press, 2004).

38   There are analogies with the work of Paul Sandby and others in their rendering of the landscapes of Highland Scotland for Ordnance Survey, where works of art acted as powerful visual records: see Stephen Daniels, *Paul Sandby: Picturing Britain* (London: Royal Academy, 2009). For more on the analogies between Sandby's art and the representation of extra-European landscapes by British artists, see John E. Crowley, *Imperial Landscapes: Britain's Global Visual Culture, 1745–1820* (London: Yale University Press, 2010), 43–5.

39   Harriet Guest, *Empire, Barbarism, and Civilisation: Captain Cook, William Hodges and the Return to the Pacific* (Cambridge: Cambridge University Press, 2007), 14.

40   For a more detailed discussion of this phenomenon, see Charlotte Klonk, *Science and the Perception of Nature: British Landscape Art in the Late Eighteenth and Early Nineteenth Centuries* (New Haven: Yale University Press, 1996).

**Figure 5.1   William Hodges, A View of the Cape of Good Hope,
taken on the spot, from on board the 'Resolution' (1772)**

Source: © National Maritime Museum, Greenwich [Photo ID: BHC 1778].

accurate coastal profile of settlements and fortifications with a discernible interest in atmospheric phenomena and climatic conditions. It was painted from Cook's cabin aboard HMS *Resolution* in November 1772, as the expedition took on provisions for its subsequent travels around the Pacific.[41] In this work, and by virtue of his innovative *plein-air* technique, Hodges encapsulated the weather conditions that ships were liable to encounter at the Cape at this time of year. The mellow hues and carefully planned compositions characteristic of Hodges's teacher, Richard Wilson, are replaced by Hodges with a strikingly realistic and atmospheric rendering of the scene before him. The shafts of sunlight breaking through the cloudy sky, the vast mass of Table Mountain, and the murky conditions through which we see the whole scene are rendered with an impressive feel for atmospheric and light conditions. Scholars have suggested that the artist's work on the voyage was influenced by his reading of Joseph Priestley's *The History*

---

41   Bernard Smith, *European Vision and the South Pacific, 1768–1850* (New Haven: Yale University Press, 1985), 57–9.

and Present State of Discoveries relating to Vision, Light and Colours (1772).[42] The use of highlights and gradations of tone combine to render the conditions unique to that moment in time. The squalls of rain and the gusts of sea-breeze that Hodges evokes in the painting actually occurred when he was preparing this work, as detailed in the ship's meteorological records kept by William Wales from 4 to 14 November 1772.[43] This approach is not some proto-Impressionist vision, concerned with capturing the fleeting atmospheric conditions of the day for purely aesthetic effect. The freshness of touch resulted from Hodges's use of the mobile studio of Cook's cabin in order to create a work intimately connected with the scientific approaches of the period. It encouraged the artist to seek out a 'typical' description of landscape; a scene that was, to use Hodges's own words, 'characteristical'.

If Hodges's work fed off the scientific and meteorological discussions that took place before and during the second voyage, then John Webber's images were part of the ethnographic bent of the third voyage. Cook describes Webber's contribution: 'Mr Webber was engaged to embark with me, for the express purpose of supplying the unavoidable imperfections of written accounts, by enabling us to preserve, and to bring home, such drawings of the most memorable scenes of our transactions, as could only be expected by a professed and skilled artist'.[44] Together with companions such as William Anderson, the botanist, and William Bayly, the astronomer, Webber contributed to the scientific achievements of the voyage. In his case, his work was instrumental in developing a wider European awareness of other people, cultures, and societies. The work of Webber, Hodges and others serves to underline the role played by art in voyages of scientific exploration and the close connections between art and science in this period.

The advent of the camera as a means of apparently objectively recording information about landscapes, wildlife and geographical features did not, as we might expect, instantly displace the expedition artist. This may have initially been due to a nervousness about the reliability and effectiveness of the new technology, which was cumbersome and erratic in its early days. Despite the potential for the difficulties inherent in the process to be exacerbated by the extreme environmental conditions, Polar expeditions were some of the earliest to take photographic equipment with them. The ill-fated 1845 voyage of John Franklin, for example, was given a camera by the photographer of the *Illustrated London News* when the camera had proved itself a useable technology only five years earlier.[45] Sadly, this expedition's photographic results are lost, although photographic portraits

42   Brian W. Richardson, *Longitude and Empire: How Captain Cook's Voyages Changed the World* (Vancouver: University of British Colombia Press, 2005), 69.

43   Smith, *European Vision and the South Pacific*, 58.

44   William Hauptmann, *Captain Cook's Painter. John Webber, 1751–1793: Pacific Voyager and Landscape Artist* (Bern: Kunstmuseum, 1996), 133.

45   Andrew Lambert, *Franklin: Tragic Hero of Polar Navigation* (London: Faber & Faber, 2009), 155.

taken of some of the men, including Franklin, before they departed, were and still are widely disseminated images. The British Arctic Expedition of 1875–76 had two photographers, George White and Thomas Mitchell, who used the latest technological processes, in addition to carrying the surgeon-artist Edward Moss. Moss's watercolours were reproduced as chromolithographs, as engravings in the press, and as magic lantern slides. As a result, they circulated widely. This overlap between science and art was evident even in the twentieth century, and again sometimes coalesced in one individual: Dr Edward Wilson, for example, surgeon and naturalist on Scott's *Discovery* and *Terra Nova* expeditions, was also a talented artist, whose watercolours complement the well-known work of expedition photographer Herbert Ponting. Ponting's beautiful landscapes have become some of the most familiar images of Antarctica. He was keen to ensure that his photography was regarded as both artistic and scientific.[46] Wilson's work, on the other hand, reminds us that Antarctica is more than a desolate land of whiteness, but rather one of often vibrant colour. His highly accurate plant and animal sketches, based on field observations, were worked up into exquisite watercolours that were works of art and, because of their accuracy and attention to detail, were used to illustrate the scientific results of the expeditions. Both photographs and works of art thus captured information and conveyed knowledge, as well as being powerful aesthetic representations of far-off lands.

Cultural Processes

In considering what constitutes a scientific instrument or instrument of exploration, one of the key corollary questions seems to be 'what is the purpose of a scientific instrument'? Is exploration solely about recording data? Or does the term allow for something else? Here we are, in some respects, questioning the definition of 'exploration' as well as the instruments used to conduct it. Is 'exploration' about filling in blanks on the map? Or about greater cultural awareness? To give a more concrete example: does collecting and bringing the world 'home' in the form of ethnographic objects constitute exploration as much as, say, recording points of latitude and longitude? About 2,000 objects from Cook voyages were extant in the nineteenth century.[47] Does 'exploration' necessarily need to be done in the field, or can it be done at home among groups of people who have never left their own

---

46 According to Max Jones, Ponting believed that his photographs should be seen alongside geological or meteorological observations as a means of knowing and understanding Antarctica: Max Jones, *The Last Great Quest: Captain Scott's Antarctic Sacrifice* (Oxford: Oxford University Press, 2003), 184. The fact that Ponting's work was exhibited at the Fine Art Society in December 1913 also speaks volumes for how he wished it to be considered.

47 Hauptmann, *Captain Cook's Painter*, 80.

shores? In other words, does ethnographic material constitute, in this context, an instrument of exploration?

Collecting and bringing artefacts back home was a crucial part of exploring, of knowing and of understanding the world beyond the usual Europe purview (although not necessarily doing so on its own terms). This is, in part, because of the process itself: in order to acquire things, people from different cultural backgrounds had to interact, understanding or misunderstanding each other as a result. Although what remains in museum collections may imply that this was a one-way process, it was not: it was only through a willingness on behalf of the people with whom they came into contact that Europeans were able to collect at all. Curiosity and an interest in the goods of the other naturally existed on both sides.[48] Objects sometimes travelled vast distances to be exchanged and could be incorporated or reshaped into new things in ingenious ways, corresponding with the first tentative steps towards mutual understanding, and a desire to incorporate the novel into existing worldviews. In preparation for a second voyage to the Pacific, Joseph Banks had replicas of Māori hand weapons, *patu*, cast in brass, ready to give as gifts. Although he did not travel on the voyage, they seem to have been given to another member of the crew on Cook's third voyage, Charles Clerke, and subsequently entered into networks of exchange on the Northwest Coast of America and beyond.[49] Thus, European-made replicas of indigenous artefact forms in novel materials made their way across the globe.

Other artefacts are material evidence of exchange in their very make-up: a shot-pouch in the NMM's collections is made from materials that had already been traded from European settlers, such as glass beads, and then made into a new artefact with design motifs that melded indigenous beliefs with patterns derived from Victorian furnishing fabrics. The material traces of the lost 1845 Franklin expedition, later characterised by the British as 'relics', were quickly incorporated into indigenous Inuit use and trade. Search expeditions retrieved these things when attempting to find out what had happened to the men; metal blades had been recycled into knives for flaying carcasses or shaping snow for igloos (NMM object AAA2098), whilst copper had been incorporated into fishhooks (NMM object AAA2302). Whilst some objects were readily adapted to local functions, others were not so straightforwardly useful – or at least something was lost in cross-cultural translation. Inuit oral histories tell of square pieces of tobacco being used by children as playthings until they got wet and began to reek: 'Smoking tobacco was far from their minds', it was said.[50] Scientific travellers did not always take

48   See, for example, Jenny Newell, 'Exotic Possessions: Polynesians and their Eighteenth-Century Collecting', *Journal of Museum Ethnography* 17 (2005): 75–88.

49   Adrienne L. Kaeppler, 'Two Polynesian Repatriation Enigmas at the Smithsonian Institution', *Journal of Museum Ethnography* 17 (2005): 152–62, quote from page 152.

50   As recounted by Rosie Iqallijuq to Dorothy Harley Eber, and quoted in *Encounters on the Passage: Inuit meet the Explorers* (Toronto, Buffalo and London: University of Toronto Press, 2008), 23. Iqallijuq had heard the stories of an elderly woman, Ulluriaq,

the trouble to try to understand the different value systems with which they were coming into contact. Rather, they took advantage of what, to them, were undeniably good deals: Dr P.C. Sutherland, who had accompanied Edward Inglefield in the *Isabel* in 1852, reported to the Ethnological Society of London that local people 'acknowledged handsome gifts of files, knives, saws and other *useful* articles, by returning with seal-skins and whalebone 50 times of more value, which they made over to us without the slightest allusion to bargain or agreement'.[51]

Some objects continued their trajectories by being brought back home. Ships had, at the very least, the capacity to ensure that more than single or singular objects were conveyed back to Britain. They could deliver large collections to museums and learned societies for further scrutiny and display, as well as returning souvenirs to loved ones and to an individual's home. Artefacts were thereby distributed widely, so facilitating their incorporation into multiple new contexts of use and display. The materiality and portability (or otherwise) of things from other cultures was important and shaped the ways in which they were viewed. There is a complex interplay between what people are prepared to relinquish and what others want to collect. Some artefacts become archetypal in the European imagination, for whatever reason: snow goggles and fishhooks are commonly found in museum collections from the Arctic, clubs and headrests from the Pacific. Their commonality may be because they were the kinds of things that were relatively easily given up and replaced, and could therefore be collected, just as much as the fact that they were felt to represent particular cultures according to European stereotypes.

The ultimate advantage of objects of exploration is their materiality, and their capacity to be moved, and seen, and, possibly, touched. Objects were used in Britain for debate or display, and contributed to the genesis and promulgation of knowledge, either about other cultures or about humankind generally. While the role of print culture in advancing ideas is well known, artefacts also circulated and were used in a similar fashion. Disciplines such as geography and anthropology (or ethnology as it was then known) were still young even by the late nineteenth century; the armchair was often the preferred site for theorists. There, scholars could digest and interpret what was brought back from voyages of exploration and advance ideas as to what it might mean. The evidence was often made to fit the theory: writing about John Lubbock, Janet Owen demonstrates how a knife collected during the 1850–55 Arctic voyage of HMS *Enterprise* was, amongst other artefacts, used for comparative purposes, drawing ethnographic objects into Lubbock's influential book, *Pre-historic Times*, with its Darwinist agenda and

---

who, according to local tradition, had been a child when William Edward Parry visited Igloolik.

51 P.C. Sutherland, 'On the Esquimaux', *Journal of the Ethnological Society of London* 4 (1856): 193–214, quote from pages 204–5.

musings on cultural evolution.[52] In some ways, as Nicholas Thomas points out, the misunderstandings or, as he puts it, rough translations of cultures and their artefacts are not as relevant as the fact that European knowledge of other areas of the world was emerging.[53]

Academic interest in the wider contexts of exploration, combined with museological and curatorial approaches, encourage us to consider a more expansive definition of the objects that contributed to the knowledge-gathering, recording and dissemination processes that were so crucial to the practices of exploration. Here again, published accounts play a crucial role. In *The Complete English Gentleman*, first published in 1730, Daniel Defoe reflected that, such was the penchant for travel literature:

> [A man] may take a tour of the world in books, he may make a master of the geography of the universe in maps, atlases and measurements of our mathematicians. He may travel by land with the historians, by sea with the navigators. He may go round the globe with Dampier and Rogers, and kno' a thousand times more doing it than all those illiterate sailors.[54]

Whatever their status as accurate accounts, Charles Batten reminds us that we should not underestimate the importance of travel accounts for well-educated people in the eighteenth century.[55]

Three-dimensional objects also illuminate the wider public engagement with exploration activity and the knowledge thus generated. In this regard, we might take the example of a papier mâché and plaster globe made by John Cary in 1791 and 'agreeable to the latest Discoveries'. It depicts the route of Cook's three voyages and carefully records his place of death. It also incorporates the track of Constantine Phipps's voyage towards the North Pole in 1773 and includes Mackenzie's explorations in Canada of 1789, thereby setting the exhibition of Pacific exploration in a broader chronological and geographical context. This object, one that could be displayed in the palm of one's hand, reinforced the idea of exploration as a distinctly British activity and one connected to rising imperial ambitions.[56] It also brought exploration beyond the confines of the scientific

---

52   Janet Owen, 'From Down House to Avebury: John Lubbock, Prehistory and Human Evolution through the Eyes of his Collection', *Notes & Records: The Royal Society Journal of the History of Science* 68 (2013): 21–34, quote from page 28.

53   Nicholas Thomas, *Islanders: The Pacific in the Age of Empire* (New Haven and London: Yale University Press, 2010), 16.

54   Richardson, *Longitude and Empire*, 146.

55   Charles L. Batten, *Pleasurable Instruction: Form and Convention in Eighteenth-Century Travel Literature* (Berkeley: University of California Press, 1978), 3.

56   National Maritime Museum, Greenwich, GLB0001. For further information, see http://collections.rmg.co.uk/collections/objects/19688.html.

academy or the Admiralty ship and into the hands of individuals who could, literally, hold the world in their hands.

Explorers and Exploration

A further question might be where do the people using these instruments of exploration – ships, paintings, but also sextants and chronometers – fit in? Does the instrument of exploration constitute the scientific object, or is it the combination of the instrument and the person doing the science or making the observations?[57] Here we might think of John Forbes Royle's inaugural lecture to King's College, London, where he reminded his audience that 'the scientific systematist, surrounded by the stores of his herbarium' should remember that the botanical specimens that formed the basis of his research had often been collected 'by adventurous and earnest men' whose role was as important in its way as his, for 'in the scientific hive as in the apiary there must be working bees and neuters as well as queens and drones: it is necessary for the economy of the commonwealth'.[58] In other words, it is not just the named 'heroes' of exploration who matter but also the less well-known members of the crew. This focus on the people that used precision instruments, not just on the objects themselves, offers the chance to touch on those personal stories beloved of museums, and to write perhaps more personal histories of exploration.

Individuals loom large in exploration histories generally, and have often been hard to escape. One need only scan the vast literature devoted to 'explorers' to appreciate the persuasive pull of personality, although recent years have been characterised by an increasing focus on other participants in such expeditions, or in less familiar aspects of them.[59] Current scholarship points to a still-strong biographical focus whilst suggesting that greater contextualisation can aid in understanding why such figures have the cultural potency they do.[60] From an

---

57 For further thoughts on the human involvement in exploration practices, see the chapter entitled 'When Human Travellers become Instruments: The Indo-British Exploration of Central Asia in the Nineteenth Century', in Kapil Raj, *Relocating Modern Science: Circulation and the Construction of Knowledge in South Asia and Europe, 1650–1900* (Basingstoke: Palgrave Macmillan, 2007), 181–222, especially 'Instruments, Travel, and Science', 202–22.

58 Endersby, *Imperial Nature*, 219.

59 For the *Terra Nova* expedition, for example, there have been biographies of Oates, Bowers, Cherry Garrard, and Crean, as well as examinations of the 'Northern Party'.

60 For example, Christina M. Adcock, Tracing Warm Lines: Northern Canadian Exploration, Knowledge, and Memory, 1905–1965, unpublished PhD thesis, University of Cambridge, 2010. The work of Felix Driver is important here: see, for example, 'The Active Life: The Explorer as Biographical Subject', *Oxford Dictionary of National Biography*, online edn, Oxford University Press, January 2006, http://www.oxforddnb.com/view/theme/94053.

artefactual point of view, museums have often found it difficult to escape from this cult of personality because of the nature of their collections. Given that places like the NMM are key public sites for the consumption of exploration histories, this is not without significance. Objects, according to the widely accepted trope, have lives of their own, but sometimes the intersection of their lives with those of particular people causes them to become entangled in a narrative that they find difficult to escape.[61] Polar exploration provides an example in the form of the 'Franklin relics'. The artefacts themselves are an eclectic jumble of fragments: buttons, spectacle lenses, slivers of soap, sledge parts and scientific instruments, to name but a few. They have, as Adriana Craciun discusses, often been difficult for British audiences to interpret, slipping between categories such as relic and commodity.[62] But within an exhibitionary perspective, what has been constant is their association with Sir John Franklin, who, as the tragic Victorian hero, has overwhelmed other participants in the 1845 expedition. From the first displays in the Royal Naval Museum's Franklin room through to the *Explorers* gallery of today, the narrative has dwelt upon 'the *Franklin* expedition' (not the British Naval Northwest Passage Expedition, as it might more properly be known), to a greater or lesser extent (Figure 5.2). Other artefacts associated with Polar exploration, but not from this expedition, have similarly become subsumed by the Franklin aura – or, at any rate, have lost any sense of a more nuanced history.[63] This is partly because of the limitations of museum display – how can a label of 50 words express the complex circulation of things?

But let us return to our focus on instruments of exploration. Human bodies clearly play an important role: scientific observations require an observer and a recorder. By its very definition, exploration necessitates 'being there', 'doing things' – explorers as subjects have to experience other places as part of their role in the creation of knowledge, and their bodies act as witnesses to this.[64] Yet this creates a central tension between objectivity and subjectivity in the practice of exploration that is key to our discussion, and that is clearly expressed both in exploration literature and in museum displays. Despite the desire of the nascent discipline of geography to systematise and regulate scientific investigations

61   For the origins of this notion, see Arjun Appadurai (ed.) *The Social Life of Things: Commodities in Cultural Perspective* (Cambridge: Cambridge University Press, 1986).

62   Adriana Craciun, 'The Franklin Relics in the Arctic Archive', *Victorian Literature and Culture* 42 (2014): 1–31.

63   See, for example, the discussion of a shot pouch collected by George Back in Claire Warrior, 'Arctic 'Relics': The Construction of History, Memory and Narratives at the National Maritime Museum', in *The Thing about Museums: Objects and Experience, Representation and Contestation* (ed.) S. Dudley et al. (London and New York: Routledge, 2012), 263–76.

64   Dorinda Outram, 'On being Perseus: New Knowledge, Dislocation and Enlightenment Exploration', in *Geography and Enlightenment* (eds) David N. Livingstone and Charles W.J. Withers (Chicago: University of Chicago Press, 1999), 281–94.

Figure 5.2   The *Explorers* **gallery at the National Maritime Museum,
demonstrating the emphasis on Sir John Franklin**

Source: © National Maritime Museum, Greenwich [Photo ID: E0077-3-2].

through the use of instruments, the subjective often still seems to win out.[65] The
NMM's Polar collections, for example, are replete with the kinds of things that
people need to support their bodies in extreme climatic conditions: clothing, food,
and specialised equipment to move around with such as sledges, snowshoes,
and snow goggles. There are also numerous artefacts connected with the mental
challenges of exploratory practice, from playbills for shipboard entertainment, to
on-board newspapers and musical instruments. In these terms, the banjo played
by Leonard Hussey to keep the spirits up of the men on Elephant Island during
Shackleton's 1914–17 Imperial Trans-Antarctic Expedition, is also an instrument

---

65   Charles W.J. Withers, 'Science, Scientific Instruments and Questions of Method
in Nineteenth-Century British Geography', *Transactions of the Institute of British
Geographers* 38 (2013): 167–78.

of exploration, essential to supporting an arduous daily existence in Antarctica (Figure 5.3). Exploration happens somewhere, and that somewhere always has an impact on people; people have to deal with basic human needs in order to do exploratory work.

The tension between the objective and the subjective has an impact on the instruments of exploration themselves. There is an interesting parallel between the way in which explorers' bodies have been seen as objective evidence of their engagement with other landscapes and environments, and the ways in which scientific instruments and other artefacts of exploration are interpreted within a museum context. In terms of provenance, this often extends to details of where things have been, even of foodstuffs: witness a ship's biscuit (ZBA5089) in the NMM's collections, which is covered with a handwritten label detailing its 'Voyage towards the North Pole Sailed 1821 returned 1823'.[66] It is not enough, then, for instruments of exploration (or biscuits) to do the work that they are sent to do – and at which they may sometimes fail. Like explorers themselves, they

Figure 5.3   Leonard Hussey's banjo

Source: © National Maritime Museum, Greenwich [Photo ID: D5239].

66   National Maritime Museum, Greenwich, ZBA5089. For further information, see http://collections.rmg.co.uk/collections/objects/563181.html.

must be made to bear witness to the geographical limits reached. Neither can we assume that instruments in and of themselves are accurate and consistently objective. Scientific instruments must be used properly to yield reliable results and are susceptible to their own failings. On the 1819–22 Overland Expedition through the Canadian Subarctic and Arctic, Captain George Back recorded that, in January 1820, the mercury had frozen in the bulb of one of the men's thermometers, probably rendering it unusable.[67] A similar story of instrumental failure can be told of the chronometers taken on William Edward Parry's 1819 voyage in HMS *Hecla*, in search of the Northwest Passage. Parry's achievement in surveying hundreds of miles of the Canadian coastline were partly due to the reliability of a Parkinson & Frodsham chronometer (ZAA0033), which continued to work in temperatures of minus 40°F, enabling him to measure accurately the expedition's longitude, although other chronometers on the voyage had failed.[68] In extreme conditions, it is not just human bodies that are vulnerable. Within practices of exploration, objects and persons are mutually implicated in the production of knowledge in complex and intriguing ways.

Conclusion

Scientific instruments are crucial to the success or otherwise of exploration. The practices of navigation, mapping, surveying, collecting, classification, and textual and visual representation that were brought to bear on the European exploration of and engagement with the wider world did not merely advance European knowledge or scientific understanding. They also worked as tools of empire by encoding cultural difference and by promoting new forms of authority over people and nature.[69] In other words, scientific instruments – and the consequences of their use – need to be situated in broader historical contexts. As we have shown, the work of scientific exploration relied on a whole assemblage of 'instruments' – ships, oil paintings, everyday objects – not immediately associated with precision technology. The material turn in histories of geography, embodied in this collection of essays, offers many exciting possibilities for future scholarship. Thinking of diverse objects in tandem, locating them individually and collectively in wider historiographical and methodological contexts, and opening up interdisciplinary connections between them and the work they performed, has the potential to bear much intellectual fruit. By bringing perspectives from histories of science, art and

---

67   C. Stuart Houston (ed.) *Arctic Artist: The Journal and Paintings of George Back, Midshipman with Franklin, 1819–22* (Montreal and Kingston: McGill-Queen's University Press, 1994), 31.

68   National Maritime Museum, Greenwich, ZAA0033. For further information, see http://collections.rmg.co.uk/collections/objects/79138.html.

69   Clayton, 'Captain Cook's Command of Knowledge and Space', 112.

empire together, such an approach promises to expand the intellectual horizons of exploration and develop our understanding of its various histories.

Chapter 6

Instruments of Exploration in National Museums Scotland

A.D. Morrison-Low

National Museums Scotland holds material relating to Scottish explorers and exploration in curatorial departments across the institution.[1] However, and in contrast, the Department of Science and Technology has relatively few major items with proven links to the history of exploration. One reason for this is that the formal collecting of historical instrumentation began much later in the Scottish national museum than it did for its sister national museums in England. Without a purchase grant of any size, and run without an influential board of trustees until 1985, the ability to acquire first-class scientific items for minimal sums of money was almost impossible: the loss of the important Sir John Findlay collection of early scientific instruments in 1961, after a loan period of 30 years, is the case that proves this point.[2] This chapter outlines elements of the institutional history of National Museums Scotland, and explains where instruments of exploration used by Scottish explorers can be found. At its heart, this chapter addresses a key matter at the heart of exploration collecting – one, indeed, that might be called the nub of curatorial anxiety – the question of collections provenance.

National Museums Scotland: A Brief Institutional History

As a national museum, National Museums Scotland has had a somewhat different remit from other London-based national museums. Much of this is due to its history and to changing collections policies.[3] Although the emphasis now, and for some time past, is on the collection of Scottish material, this is not to the exclusion of items used or originating from elsewhere. This can be summed up

---

1   I would like to thank colleagues across National Museums Scotland who have helped me with this chapter: David Forsyth, Dr Henrietta Lidchi, Neil MacLean, Julie Orford, Dr Tacye Phillipson, Dr Sarah Worden.

2   A.D. Morrison-Low, 'Sold at Sotheby's: Sir John Findlay's Cabinet and the Scottish Antiquarian Tradition', *Journal of the History of Collections* 7 (1995): 197–209.

3   G.N. Swinney, 'Towards an Historical Geography of a 'National' Museum: The Industrial Museum of Scotland, the Edinburgh Museum of Science and Art and the Royal Scottish Museum, 1854–1939', unpublished PhD thesis, University of Edinburgh, 2013.

by the phrase which, for a time, was adopted as National Museums Scotland's mission statement, and is now its 'vision statement': 'Inspiring people, connecting Scotland to the world, and the world to Scotland'.[4] Unlike the National Maritime Museum in Greenwich, established in 1936, or the National Museum of Science and Industry, a mid-nineteenth-century foundation, National Museums Scotland's collections now range across the entire spectrum of material culture and the natural world, apart from botany (dealt with in the present day by the Royal Botanic Garden Edinburgh), and the fine arts – drawings, photographs and paintings (the collected realm of the National Galleries of Scotland).[5] National Museums Scotland does, occasionally, collect examples of paintings, which are acquired for their subject matter and not because they are of any intrinsic artistic value. Thus, the only surviving portrait in oils of the young Alexander Dalrymple, who subsequently became the first Hydrographer to the Admiralty, was added to the history of science collections in 2008 (Figure 6.1). It was acquired because of the significance of who he was (and only after careful discussion with the National Galleries of Scotland). Dalrymple is shown as a young man, wearing his East India Company sea-officer's uniform, pointing at some props with a geographical flavour, including maps and globe. He is seated, probably in the Chinese drawing-room of his eldest brother's home at Newhailes, outside Edinburgh, home on leave from the Far East.[6]

The Dalrymple portrait was an exceptional acquisition for the museum's collections. Looking at a subject such as the one under consideration, 'Instruments of Exploration', collections more generally connected with exploration and geography are not necessarily to be found in the care of the Department of Science and Technology. They are more likely to reside with the Department of World Cultures, the Department of Natural Sciences, or, even, the Department of Scottish History and Archaeology. Material relating to individual Scottish explorers may well be found – as part of the dispersed national collection – at museums elsewhere around Scotland. The reasons for this lie in the way collections come together. Historically significant items become a part of the national collections through gift, loan or purchase, and strategies by individual curators may or may not work in the longer term towards their desired goals. Collecting policies held by

4   *National Museums Scotland Strategic Framework*, updated February 2011.

5   For the history of the National Maritime Museum, see Kevin Littlewood and Beverley Butler, *Of Ships and Stars: Maritime Heritage and the Founding of the National Maritime Museum, Greenwich* (London: Athlone Press, 1998); and, for the National Museum for Science and Industry, parent body of the Science Museum, see Peter J.T. Morris (ed.) *Science for the Nation: Perspectives on the History of the Science Museum* (Basingstoke: Palgrave Macmillan, 2010).

6   National Museums Scotland inv. no. T.2008.168. For Dalrymple, see Andrew S. Cook, 'Surveying the Seas: Establishing the Sea Routes to the East Indies', in *Cartographies of Travel and Navigation* (ed.) James R. Akerman (Chicago and London: University of Chicago Press, 2006), 69–96.

**Figure 6.1   Alexander Dalrymple (1737–1808), portrait in oils,
attributed to John Thomas Seton, *c.*1765**

Source: © National Museums Scotland, NMS.T.2008.168.

the institution – either publically visible or not entirely transparent – can achieve these goals. There is, however, always a random element to such additions: current individual curatorial interests or the state of the museum's purchase budget, or even a director's personal bias can always militate against particular acquisitions.

As a short introduction to the institutional history of National Museums Scotland (and thus to its collections), today's buildings and collections are the product of the fairly-recent amalgamation of two national museums. One was the former museum of the Society of Antiquaries of Scotland which was founded in 1780 as a manifestation of that interest in natural products and cultural artefacts in the Enlightenment in Scotland.[7] The other was a government-sponsored body, initially established as the Industrial Museum of Scotland, set up from the proceeds of the Great Exhibition – as were sister institutions in South Kensington and Dublin – in 1854.[8] The technological interest was associated with this second strand, which was subsequently joined by a third, when the natural history collections (including some important ethnographic material) of the University of Edinburgh were added to the Chambers Street site in the late nineteenth century. By this time, the Industrial Museum of Scotland had entered a second phase as the Edinburgh Museum of Science and Art, with the emphasis on the second of those disciplines. In 1901, Edward VII was inclined to allow the museum to be renamed the 'Royal Scottish Museum', the same year that the new Technology Department gained staff and a separate identity. By and large, the museum has remained that way, although it recently shed the 'Royal' label. Until recently, the displays were shown in discrete single-discipline galleries. This, too, has changed.

The National Museums of Scotland was formed in 1985 by the amalgamation under one director of the National Museum of Antiquities of Scotland and the Royal Scottish Museum. The collections of the Antiquities Museum, telling the story of Scotland from early archaeological times until the twentieth century, have been moved into the newish wing (the Museum of Scotland) attached to the end of the former Royal Scottish Museum. This was renamed from 1985 the Royal Museum of Scotland, and, since its re-opening in 2011, the National Museum of Scotland. The new Museum of Scotland was opened by Her Majesty the Queen in 1998, and remained open while the Victorian Royal Museum, refurbished in a £46.9 million makeover, re-opened in summer 2011.[9]

How Historic Instrument Collections are Put Together

As hinted at above, collections may be formed at a very much more detailed level than the museums that contain them. Some history of science items came into the collections during the early days of both the National Museum of Antiquities and the Industrial Museum of Scotland, albeit in a rather ad hoc manner. For example, an

---

7   Alan S. Bell (ed.) *The Scottish Antiquarian Tradition: Essays to Mark the Bicentenary of the Society of Antiquaries of Scotland and its Museum, 1780–1980* (Edinburgh: John Donald, 1981).

8   Jenni Calder, *The Royal Scottish Museum: The Early Years* (Edinburgh: Royal Scottish Museum, 1984).

9   Jenni Calder, *The Wealth of a Nation in the National Museums of Scotland* (Edinburgh: National Museums of Scotland, 1989); Charles McKean, *The Making of the Museum of Scotland* (Edinburgh: NMS Publishing, 2000); Helen Smailes, *A Portrait Gallery for Scotland : the Foundation, Architecture and Mural Decoration of the Scottish National Portrait Gallery, 1882–1906* (Edinburgh: National Galleries of Scotland, 1985); Duncan Thomson, *A History of the Scottish National Portrait Gallery* (Edinburgh: National Galleries of Scotland, 2011).

early seventeenth-century English tripod microscope was presented to the Society of Antiquaries of Scotland before 1892.[10] An important array of chemical glassware, probably made in Leith and used in lecture-demonstrations at the University of Edinburgh by the Enlightenment chemist and natural philosopher Joseph Black, came to the then Industrial Museum of Scotland in 1858. This material was presented by the incoming professor of chemistry, Lyon Playfair, to his colleague and friend, the new director of the Industrial Museum of Scotland, George Wilson, and is part of a group of early chemical equipment known by its donor's name as the Playfair Collection. Black's international reputation rests on his discovery of carbon dioxide, and development of the concepts of specific and latent heat.[11] Neither of these acquisitions would have been considered of any significant monetary value at the time of their entry into the museums' collections, although they would have been seen as important in terms of material culture. They have subsequently gained in importance in terms of material culture through display and publication.

Although other historic instruments came into the collections after these small beginnings, National Museums Scotland has been collecting scientific material in a structured way since only 1968. This is when a curator, David Bryden, was specifically appointed to do so. The discipline of instrument history was itself then also young, and after presenting a paper on the subject to the British Society of the History of Science, Bryden authored a publication that was as much a road map for future work as a document identifying sources for uncovering the roots of this activity. Bryden's *Scottish Scientific Instrument-Makers 1600–1900* (1972) provided the basis for another, rather more substantial, foray into the subject which approached the history of technology through business histories: *Brass & Glass: Scientific Instrument Making Workshops in Scotland*, published in 1989. This, in turn, is now due for a reassessment.

Part of the curatorial task is to publicise material that comes into the collection, ensuring that knowledge about the material can inform future acquisition. George Wilson, for instance, wrote about the history of the air pump, and was prompted to do so by the acquisition of an early eighteenth-century Hauksbee-type example, 'rescued from the lumber room of the laboratory' along with chemical glassware.[12]

---

10   G.L'E. Turner, 'Decorative Tooling on 17th and 18th-century Microscopes and Telescopes', *Physis:Rivista Internazionale di Storia della Scienza* 8 (1966): 99–128.

11   Provenance and method of acquisition of this material is discussed by R.G.W. Anderson, *The Playfair Collection and the Teaching of Chemistry at the University of Edinburgh, 1713–1858* (Edinburgh: Royal Scottish Museum, 1978), 135–47; and by A.D. Morrison-Low, 'Surviving Eighteenth-Century Chemical Apparatus in the National Museums of Scotland', in *Edinburgh 300: Cradle of Chemistry: The Early Years of Chemistry at the University of Edinburgh* (ed.) Robert G.W. Anderson (Edinburgh: John Donald, 2015), 131–8.

12   Lyon Playfair, *A Century of Chemistry in the University of Edinburgh: being the Introductory Lecture to the Course of Chemistry in 1858* (Edinburgh: Murray & Gibb, 1858), 13, quoted in Anderson, *The Playfair Collection and the Teaching of Chemistry at the University of Edinburgh, 1713–1858*, 59; George Wilson, 'On the Early History of the

It must be emphasised that the title of this chapter, 'instruments of exploration' is problematic for National Museums Scotland, as much of the historic instrumental material is without an easily identifiable provenance. It does, however, provide an opportunity to discuss why this should be, and to investigate the nature of the importance of provenance generally. In the past, items were acquired as good examples of particular types of device that illustrated a development in the evolution of that instrument. Thus, 'gaps in the collection' were filled by following identifiable technological histories of particular apparatus. In 1979, National Museums Scotland acquired 58 microscopes which dated in their manufacture from between 1800 and 1860. The group of instruments was acquired because it showed the construction and especially the optical improvements to the lens system of the instrument. This incoming material, as a group, dovetailed with almost no duplication items already in the collection (an important consideration, as doubling of similar items is expensive in terms of storage space and conservation time): but all 58 pieces were without provenance.[13]

The collecting of instruments, as with any other form of proactive collecting (as opposed to the passive collection of items offered by gift or bequest), depends to a large extent on what is available on the open market at any given time; what the museum's current collecting policy is; and whether or not the museum's purchase grant is in a healthy condition. Even so, individual curators, who should be alert about what is already within their area of responsibility, have to convince an ascending hierarchy of colleagues that obtaining a piece is either a reasonable outlay of taxpayers' money, or, if (increasingly unlikely) the gift is worthy of a place in perpetuity in the nation's collections, with subsequent likely expenditure on its storage and conservation. Other factors, such as any given director's personal scholarly interest, may mean that items recommended by individual curators are rejected in favour of items that 'fit' elsewhere.

Provenance and Instrumentation

Perhaps the most significant item within National Museum Scotland's collections of historic scientific instruments to be used on a voyage of discovery was one taken on Captain James Cook's first Pacific voyage of 1768 to 1771. This expedition had the twin objectives of observing the 1769 Transit of Venus across the Sun's disc, and to uncover positive evidence of the 'unknown southern continent'. The instrument of exploration in question is an astronomical regulator (Figure 6.2). This astronomical

---

Air–Pump in England', *Edinburgh New Philosophical Journal* 46 (1849): 330–54. The example in National Museums Scotland has the identifying number T.1858.275.1. It has been most recently assessed by Terje Brundtland, 'Francis Hauksbee and his Air Pump', *Notes and Records of the Royal Society* 66 (2012): 253–72.

13   See R.H. Nuttall, *Microscopes from the Frank Collection 1800–1860* (Jersey: Arthur Frank, 1979).

Figure 6.2   Astronomical regulator by John Shelton, 1756
Source: © National Museums Scotland, NMS.T.1978.1.

regulator, made by John Shelton of London, has had its provenance uncovered by careful research published in 1969 by Derek Howse and Beresford Hutchinson, then both of the National Maritime Museum. This was their final article in a series discussing the history of Captain James Cook's clocks and watches. The regulator came to the Royal Scottish Museum in 1978 from the Edinburgh Meteorological Office. As reported by Howse and Hutchinson, the initials 'KO' (probably for 'Kew Observatory') were found scratched below Shelton's signature. Howse and Hutchinson were thus able to match it to a clock on the Royal Society's inventory. Subsequently, the regulator appears to have travelled with Sir Thomas Brisbane to Parramatta Observatory, New South Wales in 1822–23, and then, from London, with Captain Henry Foster, RN, in HMS *Chanticleer* around South America and South Africa. The inscription on the back of the hour dial 'J[ames] Fayrer, Royal Observatory Cape of Good Hope, July 27th 1829/Cleaned &c.', pins it to a particular time and place. But the decisive evidence was finding the words 'J[ohn] Shelton October ye 8 1756' near Fayrer's inscription. In combination with other evidence, this wording strongly suggested that it was the clock bought by the Royal Society for 30 guineas in 1760. After going to St Helena, the Cape of Good Hope, Barbados, and being used to set out the Mason-Dixon line, the regulator travelled with Cook between 1768 and 1771 in the *Endeavour* for the Transit of Venus, and was then used by Neville Maskelyne at Schiehallion in Perthshire, in experiments to estimate the weight of the Earth.[14]

This astronomical regulator is the most fully-researched instrument connected with exploration in the National Museums Scotland collections. Even those pieces – sextants, octants, artificial horizons and other nautical and hydrographic material – which came to the museum from the Admiralty in 1910 and 1911, with information actually engraved upon them, still need further research, most probably in the Admiralty records held in the National Archives at Kew, or in the Hydrographic Office at Taunton. These instruments had clearly been used over the previous century by the Royal Navy in its peace-keeping role, while mapping and surveying the oceans of the world in a time of relative peace. For example, there is a fine example of a repeating circle, made in about 1800 by Thomas Jones, a scientific instrument maker based in London's Charing Cross and an apprentice to the eminent maker, Edward Troughton (Figure 6.3). A repeating circle is similar to the reflecting circle, in which a series of observations are 'repeated', so that the error between readings is increasingly narrowed. This portable instrument, intended for astronomical readings (possibly on a voyage of exploration), was derived from one described by Edward Troughton in 1822. In 1985, Alan Stimson of the National Maritime Museum identified the numbering system proposed in 1828 by the Hydrographer of the Navy, Sir William Parry. This was put into effect by Thomas Jones, who held the instruments made for the Hydrographic Office from early 1826. In February 1828 these instruments were all returned to the Admiralty and marked with various letters. Stimson gleaned from early Hydrographic Office letter books the numbering system: A for perambulators and transit instruments; B for theodolites; C for altitude and azimuth instruments, and so on.[15] This particular repeating circle has the Admiralty mark 'C 1', which might therefore identify it with a particular voyage.[16]

---

14   Derek Howse and Beresford Hutchinson, 'The Saga of the Shelton Clocks', *Antiquarian Horology* 6 (1969): 281–98. For Maskelyne on Schiehallion, see Derek Howse, *Neville Maskelyne: The Seaman's Astronomer* (Cambridge: Cambridge University Press, 1989); and Nicky Reeves, '"To Demonstrate the Exactness of the Instrument": Mountainside Trials of Precision in Scotland, 1774', *Science in Context* 22 (2009): 323–40.

15   Alan Stimson, 'Some Board of Longitude Instruments in the Nineteenth Century', in *Nineteenth-Century Scientific Instruments and their Makers* (ed.) P.R. de Clercq (Amsterdam; Atlantic Highlands, NJ; Leiden: Distributed in the USA by Humanities Press: Museum Boerhaave: Rodopi, 1985): 93–115, and see especially pages 114–15.

16   NMS.T.1910.89. Since this paper was delivered in May 2012, Megan Barford has undertaken research towards her PhD at the University of Cambridge, using material from the Hydrographic Office at Taunton which has revealed that 'C1' was used in Australia in 1842 by Phillip Parker King, the implication being that the circle was taken to Australia in 1832, then transferred to Captain Francis Price Blackwood for his survey work in HMS *Fly* on the north east coast of Australia in 1842. I am grateful to her for this information.

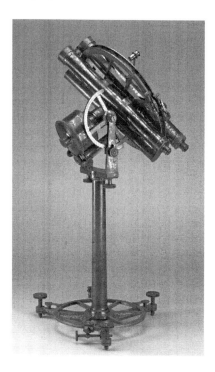

**Figure 6.3   Repeating circle by Thomas Jones, marked with
                  the Admiralty mark 'C 1', *c*.1800**

Source: © National Museums Scotland, NMS.T.1910.89.

Other pieces from the same collection give clues as to their ultimate destination: for instance, a new deep-sea self-registering thermometer was patented by the London instrument makers Negretti & Zambra in 1874 that contained only mercury and had no indexes, unlike the Six thermometer, previously in use. It was made in both straight and U-forms.[17] Examples were sent for trials aboard HMS *Challenger*, and when, in 1875, her captain, George Nares, was detached from *Challenger* to lead the Arctic expedition in HMS *Alert*, he took some of the new apparatus with him, including an example which is marked with a government arrow, the number '17' and 'Polar Expedition 1875'. This was part of the donation to the Museum

---

17   Anita McConnell, *Historical Instruments in Oceanography: Background to the Oceanography Collection at the Science Museum* (London: HMSO, 1981), 34; Anita McConnell, *No Sea Too Deep: the History of Oceanographic Instruments* (Bristol: Adam Hilger, 1982), 99–100.

from the Admiralty in 1911.[18] In 1910 and 1911, the Hydrographic Department was disposing of its redundant equipment and this was divided between the Royal Scottish Museum and the Science Museum in London, this being before the inception of a national maritime museum which had to wait until 1936.[19] The eminent Dutch maritime scholar, Dr Willem Mörzer Bruyns, has recently reported upon the items in the navigation instrument collection, commenting that:

> A significant acquisition was in 1910 and 1911, when the Admiralty donated 28 objects from the Hydrographic Office, including an octant, several sextants, artificial horizons, and a dip sector. Three of the artificial horizons and the dip sector have a stuck-on MS label on their boxes, stating that the Admiralty had lent them to an exhibition that was specially organized for the Sixth International Geographical Congress, held in London in 1895. Some of the Hydrographic Office instruments, including two aluminium-frame quintants by Cary of London, are inscribed "Ant: Ex: 1901"; they were taken – or intended to be taken – on the British Antarctic Expedition of 1901–4, the Discovery Expedition under Captain Robert Falcon Scott, RN.[20]

HMS *Challenger* and Edinburgh's Contribution to Oceanography

Exploration by the Royal Navy was not the only form of geographical investigation during the nineteenth century. Edinburgh's involvement in the new science of oceanography came towards the end of the nineteenth century, although it had earlier roots: William Scoresby, Robert Jameson, and Edward Forbes are the three distinguished figures in this respect as pioneers of this then-infant science.[21] Dr Tony Rice has credited the mid-Victorian naturalist W.B. Carpenter as being the chief instigator of the *Challenger* Expedition. By 1869, Carpenter was

18   NMS.T.1911.62. Discussed by A.D. Morrison-Low, 'Marine Sampling and Collecting Equipment in the Collections of the National Museums of Scotland', *Scottish Naturalist* 111 (1999): 137–55.

19   Willem F.J. Mörzer Bruyns, 'Navigational Instruments in the Collection of the Science Museum, London: A Report', *Bulletin of the Scientific Instrument Society* 108 (2011): 34–43.

20   Willem F.J. Mörzer Bruyns, 'Collection of Navigational Instruments at the National Museums Scotland', *Bulletin of the Scientific Instrument Society* 111 (2011): 17–25. The aluminium instruments are NMS.T.1911.34 and .35.

21   For Scoresby, see Anita McConnell, 'The Scientific Life of William Scoresby Jnr., with a Catalogue of his Instruments and Apparatus in the Whitby Museum', *Annals of Science* 43 (1986): 257–86. For Jameson, see Jessie M. Sweet, 'Robert Jameson and Shetland: a Family History', *Scottish Genealogist* 16 (1969): 1–18; Jessie M. Sweet, 'Robert Jameson and the Explorers: the Search for the North-West Passage Part 1', *Annals of Science* 31 (1974): 21–47. For Edward Forbes, see E.L. Mills, 'A View of Edward Forbes, Naturalist', *Archives of Natural History* 11 (1982–4): 365–93.

convinced as a result of his personal participation on the recent successful voyages of the *Lightning* and the *Porcupine* that the British Government should fund a major scientific circumnavigation. Through his scientific contacts (he was a Vice-President of the Royal Society), and some personal political lobbying (he was acquainted with the Prime Minister, William Gladstone), Carpenter managed to raise the money and set up the structure for the expedition: he was perhaps assisted by a recognition by the Hydrographic Department of the pressing necessity of investigation of the ocean depths for submarine telegraphy.[22] The *Challenger* expedition, a joint venture between the Admiralty and the Royal Society, appointed as its scientific director Charles Wyville Thomson, Professor of Natural History at the University of Edinburgh, and Keeper of Natural History in the Edinburgh Museum of Science and Art (a precursor of the Royal Scottish Museum, and thus, National Museums Scotland), who had gained zoological experience from earlier expeditions on the *Lightning* and the *Porcupine*.[23] Carpenter, aged 59 by 1872, was dissuaded from again going to sea. It is not necessary to do more here than outline *Challenger*'s achievements: with her scientific staff of six, besides the ship's officers and crew of 240, the wooden converted steam-assisted corvette covered almost 70,000 miles of the world's oceans between 1872 and 1876. Surveys were made, collections amassed and magnetic and meteorological registers kept. On the expedition's return the scientific results were published in 50 volumes between 1880 and 1895: 715 new genera and 4,417 new species were identified, and a mass of accumulated information about the physics, chemistry and geology of the land and sea was presented.[24]

The bulk of *Challenger*'s collections eventually went to the Natural History Museum in London. Forewarned by the dismal fate of the specimens from earlier voyages of exploration, Wyville Thomson had managed to arrange that material gathered on the *Challenger* expedition should become government property, and that funding should be available to publish results undertaken through his personal supervision, in Edinburgh.[25] Small numbers of natural history specimens from the voyage have found a home in Edinburgh. Amongst them are 171 duplicate

---

22 A.L. Rice, *British Oceanographic Vessels, 1800–1950* (London: Ray Society, 1986), 30; Roger Smith, 'Carpenter, William Benjamin (1813–1885)', *Oxford Dictionary of National Biography*, Oxford University Press, 2004; online edn, May 2006 (http://www.oxforddnb.com/view/article/4742).

23 For C. Wyville Thomson, see William A. Herdman, *Founders of Oceanography and Their Work: an Introduction to the Science of the Sea* (London: Edward Arnold, 1923), 37–68; Margaret Deacon, *Scientists and the Sea, 1650–1900: A Study of Marine Science* (New York and London: Academic Press, 1971), 333–406.

24 *Report on the Scientific Results of the Voyage of H.M.S.* Challenger *During ... 1873–76*, 50 volumes (London: John Murray, 1880–1895); for a synopsis, see Eric Linklater, *The Voyage of the 'Challenger'* (London: Cardinal, 1974), and Deacon, *Scientists and the Sea, 1650–1900: A Study of Marine Science*, 333–65.

25 This unedifying story is recounted by Deacon, *Scientists and the Sea, 1650–1900: a Study of Marine Science*, 366–72.

(that is, gathered at the same place by the same collector as the type specimen) Echinoderms, Tunicates, Crustacea, and Sponges which were presented by the Natural History Museum in London in 1899 to the Natural History department in the museum in Edinburgh, and a giant clam, that arrived a century later.[26] When the centenary exhibition of the great oceanographic voyage was held in the Royal Scottish Museum in 1972 there was some difficulty in gathering material together: there was, for instance, no contemporary model of the famous vessel herself, and one had to be commissioned.[27]

Of all the instruments held in National Museums Scotland which illustrate the *Challenger* voyage, the ones with the strongest connection to the voyage's pioneering work in oceanography are the six Miller-Casella deep-sea thermometers with protective housing. These were amongst the 30 retested – many of them to destruction, hence their less-than-pristine condition – by P.G. Tait, Professor of Natural Philosophy at the University of Edinburgh after *Challenger*'s return.[28] There are also examples of the hydrometers used by the marine scientist, J.Y. Buchanan, in on-board experiments to measure the salinity of seawater at different stations.[29]

The *Scotia* Expedition, 1902–1904

The close associations between Edinburgh University and the *Challenger* expedition turned the attention of William Spiers Bruce to the sea and exploration. He was admitted to the University of Edinburgh to read medicine, but he did not graduate. In 1892–93 Bruce sailed as ship's surgeon in the Dundee whaler *Balaena* for the South Shetlands (between the Falkland Islands and Antarctica) where he undertook zoological research and discovered the first evidence of the Antarctica anticyclone. After numerous fruitless attempts to raise funds for the exploration

---

26   'Appendix: Report of the Director of the Edinburgh Museum of Science and Art for 1899', in *Forty-Seventh Annual Report of the Department of Science and Art for 1899: HL Deb 25 May 1900 vol 83 c1227*, 7 and 9. The giant clam, MS.Z.2002.156, was brought back from Australia by Sir John Murray as a souvenir. His grand-daughter Miss Rhoda Murray inherited it and lent the clam to the Museum for its Challenger exhibition in the 1970s.

27   NMS.T.1972.95. Made by H.J. Boyd of Edinburgh, and discussed by J.D. Storer, *Ship Models in the Royal Scottish Museum, Edinburgh: a Catalogue of Models Representing the History of Shipping from 1500 BC to the Present Day* (Edinburgh: Royal Scottish Museum, 1985), 43–4.

28   NMS.T.1999.382.1-.7. P.G. Tait, 'The Pressure Errors of the Challenger Thermometers', *Challenger Narrative*, vol. 2, Appendix A.

29   NMS.T.1922.3.1-.3 and T.1922.4.1-.3. For a recent assessment of J. Y. Buchanan, an important figure in Scottish oceanography, see Margaret Deacon, 'Buchanan, John Young (1844–1925)', *Oxford Dictionary of National Biography*, Oxford University Press, May 2006 (http://www.oxforddnb.com/view/article/46611)

of South Georgia, he took charge of the Ben Nevis Meteorological Observatory in 1895, but left the following year to pursue biological research on an expedition to Franz Joseph Land.[30]

Over the next few years Bruce went on several expeditions, as the result of which he gained experience in polar work. In 1900 he announced his plans for a Scottish National Antarctic expedition to explore the Weddell Sea. Funds were raised in Scotland, and the *Scotia*, a converted Norwegian whaler, left the Clyde in November 1902. She spent two summers in oceanographic work in the Weddell Sea and the South Atlantic, returning to Scotland in July 1904 with large biological collections from depths as great as 2,900 fathoms and voluminous hydrographical and meteorological observations. The results of the expedition were published in seven volumes, but the series was left incomplete because of lack of funds.[31] Bruce went on several other polar expeditions, and, in 1907, founded the Scottish Oceanographical Laboratory, hoping that it would flourish and grow to become a great institute of oceanography comparable to that of Monaco. He undertook the burden of the Laboratory's upkeep, but failing health and lack of money compelled him to abandon the scheme in 1920. The bulk of the natural history specimens and much of the residue of the physical apparatus of the Scottish Oceanographical Laboratory came to the Royal Scottish Museum in the following year; how much of the hardware was actually used at sea is unclear. Other natural history specimens were presented to museums in Aberdeen, Glasgow, Paisley, Perth and Dundee, while the University of Edinburgh and the Royal Scottish Geographical Society received manuscripts, maps, and photographs.[32]

The specimen of the Adélie penguin *Pygoscelis adeliae*, which was brought back amongst the *Scotia*'s collections, and the now famous photograph of the piper with the penguin are perhaps among the more memorable items connected with the Scottish Antarctic Expedition of 1902–1904.[33] As striking items, both were displayed on the 'Instruments of Science' gallery in the National Museums of Scotland until its closure in 2008. Less dramatic display items, also in the same gallery, were two thermometers marked 'SCOT. ANT. EXPED. 1902', one with a

30   On Bruce, see Peter Speak, *William Speirs Bruce: Polar Explorer and Scottish Nationalist* (Edinburgh: NMS Publishing, 2003).

31   R.N. Rudmose Brown, *The Voyage of the 'Scotia': Being the Record of a Voyage of Exploration in Antarctic Seas* (Edinburgh: William Blackwood and Sons, 1906); *Scottish National Antarctic Expedition: Report of the Scientific Results of the Voyage of S.Y. 'Scotia' During the Years 1902, 1903 and 1904*, 7 volumes (Edinburgh: Scottish Oceanographical Laboratory, 1907).

32   G.N. Swinney, 'Some New Perspectives on the Life of William Speirs Bruce (1867–1921): with a Preliminary Catalogue of the Bruce Collection of Manuscripts in the University of Edinburgh', *Archives of Natural History* 28 (2001): 285–311; G.N. Swinney, 'William Speirs Bruce, the Ben Nevis Observatory and Antarctic Meteorology', *Scottish Geographical Journal* 118 (2002): 263–82.

33   G.N. Swinney, 'One Hundred Years Ago and Pipers Head for the Antarctic', *Piping Times* 55 (2002): 29–33.

mercury indicator, the other with a spirit indicator, both constructed by Negretti & Zambra, and presented to the museum by the Admiralty as part of that off-loading of its redundant apparatus in 1911 discussed above.[34] The museum continues to acquire items associated with oceanographic expeditions when (and if) they become available: a sextant, acquired in 1981, was originally retailed in Liverpool, and has a note in it to say that it was used by Jock Fitchie, mate on the *Scotia* when it went to the South Polar regions on Bruce's Scottish National Antarctic Expedition.[35]

Provenance and Explorers on Land

It has to be admitted, nevertheless, that when it comes to instruments used by famous Scottish explorers on land, the Scottish national collections under-represent the land-based exploratory endeavours of numerous men and women. A few years ago, colleagues were keen to illustrate the Scottish Diaspora and asked for material to help illustrate the work of Alexander Mackenzie, of Canadian exploration fame, with instruments of the type perhaps used by the explorer in his discovery of the Mackenzie River in Canada's north-west territories during the 1790s. It was difficult to know, however, exactly what these might have been since Mackenzie never mentioned them in his writings and apparently none has survived. It is possible that he travelled and navigated without the aid of instruments, at least initially. Mackenzie wrote:

> In this [first] voyage, I was not only without the necessary books and instruments, but also felt myself deficient in the sciences of astronomy and navigation: I did not hesitate, therefore, to undertake a winter's voyage to this country [i.e. England], in order to procure the one and acquire the other. These objects being accomplished, I returned, to determine the practicability of a commercial communication through the continent of North America, between the Atlantic and Pacific Oceans ... Nor do I hesitate to declare my decided opinion, that very great and essential advantages may be derived by extending our trade from one sea to another.[36]

His purpose was to reach the Pacific Ocean overland, in order to open up and exploit new territories for the fur trade, under the auspices of one of the aggressively commercial Canadian fur companies, the North West Company. It is not know from whom Mackenzie acquired his instruments, and who gave him instruction in

---

34   NMS.T.1911.49 and .50.

35   NMS.T.1981.59.

36   Alexander Mackenzie, *Voyages from Montreal: on the River St. Laurence, through the Continent of North America to the Frozen and Pacific Oceans in the Years 1789 and 1793; with a Preliminary Account of the Rise, Progress and Present State of the Fur Trade of that Country* (Edinburgh: Printed for T. Cadell, jun. and W. Davies ... and W. Creech, 1801), Preface, v.

their use, but it is safe to surmise that both were based in London rather than in his native Stornoway. Contact with colleagues in the Canada Science and Technology Museum in Ottawa provided more information about what these instruments may have been, but the instruments themselves do not appear to have survived either in Canada or in Scotland.[37]

Similarly, although National Museums Scotland holds two scientific items known to be connected with Sir Thomas Makdougall Brisbane – who at one point used the above-mentioned Shelton regulator at his observatory at Parramatta in New South Wales in the 1820s – there are more items from his collections to be found in the Department of Natural Sciences or the Department of World Cultures than in the Department of Science and Technology.[38] Scientific instruments were manifestly important to Brisbane. Not only did he found an astronomical observatory in the southern hemisphere, explicitly to map the uncharted heavens, but he was also a hands-on astronomer himself. Brisbane, a career soldier, had at one point found himself on his way, with his regiment in a commandeered Newcastle collier, across the Atlantic to a theatre of war in the West Indies in 1795. An error in longitude made by the crew, unused to deep sea navigation, resulted in their almost being shipwrecked. In Brisbane's own words:

> After our vessel had sailed alone for some weeks, the mate came to my cabin one morning at 4 o'clock and awoke me, to say that they had made the land; but he was afraid it was the main continent. I immediately got upon deck and found the ship among the breakers; and the captain on seeing the danger, said, – "Lord have mercy on us, for we are all gone". I said that is all very well, but let us do everything we can to save the ship. He ordered the helm to be put hard down; but so completely were the seamen paralysed by their awful situation, that not one of them would touch a rope. With the assistance of the officers, I, with my own hands, eased off the main-boom to allow the ship to pay off, and the sail to draw upon the other tack. Most providentially the wind came from the coast, and filled the sails, and though we were from four till ten in the morning in this critical juncture, yet we found ourselves at length off the bank.

> Reflecting that I might often, in the course of my life and services, be exposed to similar errors, I was determined to make myself acquainted with navigation and nautical astronomy; and for that purpose, I got the best books and instruments, and in time became so well acquainted with these sciences, that when I was returning home I was enabled to work the ship's way; and having since crossed the tropics eleven times, and circumnavigated the globe, I have found the greatest

---

37   Frank C. Swannell, 'Alexander Mackenzie as Surveyor', *The Beaver* (Winter 1959) [no volume]: 20–25.

38   For Brisbane's collections in National Museums Scotland, see A.D. Morrison-Low, 'The Soldier-Astronomer in Scotland: Thomas Makdougall Brisbane's Scientific Work in the Northern Hemisphere', *Historical Records of Australian Science* 15 (2004): 151–76.

> possible advantage from my knowledge of lunar observations and calculations
> of the longitude. In proof of which, in sailing from Port Jackson [or Sydney] to
> Cape Horn in 1825, a distance of about 8000 miles, I predicted our making the
> land to within a few minutes.[39]

Where are Brisbane's instruments now? He is justifiably famous for building and equipping the first astronomical observatory in Australia, for which he shipped his own high specification London-made items from his Largs observatory along with his astronomer, James Dunlop, to New South Wales. Many of the instruments remain in Australia today, although the observatory itself no longer survives.[40]

With respect to instruments relating to the exploration of Scotland in earlier periods, we know that Scotland was first mapped at the end of the sixteenth century and in the early years of the seventeenth century by Timothy Pont, who travelled the length and breadth of the kingdom undertaking what has been seen as chorographical survey: in the terms of this chapter, he can be seen as an explorer of his own nation. He died before he completed the task, and King James VI gave instructions that his manuscript maps should be purchased from his heirs and prepared for publication, but thanks to the civil unrest of the time they were nearly forgotten, and some were lost or destroyed. Robert Gordon of Straloch was prevailed upon to undertake their revision with a view to subsequent publication: this included Gordon undertaking new surveys himself. Robert Gordon's amendments and additions were completed by his son, James Gordon, and published in the fifth volume of Joan Blaeu's *Atlas Novus* in Amsterdam in 1654. Scotland was then reckoned to be the most comprehensively mapped country in Europe.[41]

Nothing is known of Pont's instrumentation; but an instrument connected with Robert Gordon survives. This is a French Gothic astrolabe, made in the early fifteenth century in the workshop of Jean Fusoris of Paris. None of Fusoris's astrolabes is signed, but of the 18 that survive, all have the same characteristic error: the presence of the star Cornu Arietis (β Arietis) in the southern, instead

---

39   [Thomas Makdougall Brisbane], *Reminiscences of General Sir Thomas Makdougall Brisbane* (Edinburgh: Thomas Constable, 1860), 13–14.

40   Shirley D. Saunders, 'Sir Thomas Brisbane's Legacy to Colonial Science: Colonial Astronomy at the Parramatta Observatory, 1822–1848', *Historical Records of Australian Science* 15 (2004): 177–209; Simon Schaffer, 'Keeping the Books at Paramatta Observatory', in *The Heavens on Earth: Observatories and Astronomy in Nineteenth-Century Science and Culture* (ed.) David Aubin, Charlotte Bigg and H. Otto Sibum (Durham and London: Duke University Press, 2010): 118–47. Surviving instruments are to be found at the Powerhouse Museum Sydney, and can be seen online at http://www.powerhousemuseum.com/collection/database/theme,798,Parramatta_Observatory_-_Australias_First_Observatory.

41   Ian C. Cunningham ed., *The Nation Survey'd: Essays on Late Sixteenth-Century Scotland as depicted by Timothy Pont* (East Linton: Tuckwell, 2001); Jeffrey Stone, *The Pont Manuscript Maps of Scotland: Sixteenth Century Origins of a Blaeu Atlas* (Tring: Map Collector Publications, 1989).

of the northern, hemisphere. In the late 1590s, Robert Gordon went to Paris to complete his education. It is now thought that this is when Gordon acquired this instrument, which was used by surveyors to solve problems of time, position, and trigonometry. It has had some alterations and additions made to it. It is inscribed around the rim 'Robertus Gordonius', and on the back below the shadow square with the date '1597'. It has also had two projection plates re-engraved, one for the latitude of Edinburgh, the other for Straloch, in Aberdeenshire.[42] This instrument emerged again during the mid-nineteenth century, when it was first lent by the artist Horatio McCulloch, RSA, to an Edinburgh exhibition of Industrial and Decorative Art in 1861, and then probably sold after his death (but poorly described) in the sale of his effects in 1867.[43] The item then disappeared from view until 1927, when A.S. Cumming reported that he had found it in an antiques shop; his widow subsequently bequeathed it to the Royal Scottish Museum in 1947.[44]

Timothy Pont may have explored his native country at a period when travel within it was hazardous, but who should be judged as a Scottish explorer as opposed to an explorer of Scotland? They tend to be later figures. Mungo Park, for instance, can be counted as a famous Scots-born explorer of Africa who attempted to solve at first-hand the two key geographical questions directed at the River Niger in the late eighteenth century: whether it flowed from east to west, or from west to east, and where it finally emptied. There were several contradictory theories about this.[45] Park drowned in 1806 during his second journey: his door plate, dating from the time when he was a surgeon in Peebles in the Scottish Borders, is now in the National Museums' collection.[46]

---

42 NMS.T.1947.27: Angus Macdonald and A.D. Morrison-Low, *A Heavenly Library: Treasures from the Royal Observatory's Crawford Collection* (Edinburgh: Royal Observatory and National Museums of Scotland, 1994), 25.

43 *Board of Manufactures. Official Catalogue of the Exhibition of Industrial and Decorative Art, 1861, in the National Gallery Buildings* (Edinburgh: HMSO, 1861), item 577: 'a brass astrolabe, [signed] Robertus Gordinus, 1597'; *Catalogue of the ... Collection ... of the late Horatio McCulloch Esq., ... Sold by Auction by Mr T. Chapman* (Edinburgh, 28 November 1867): discussed by A.D. Morrison-Low, 'Sold at Sotheby's' (n. 2), quote at page 200.

44 A. S. Cumming, 'Gordon of Straloch's Astrolabe', *Scottish Geographical Magazine* 42 (1926): 79–82. Regarding his acquisition of the piece, Cumming wrote how 'Many years ago, in an antique-furniture shop in Edinburgh, I came across an old astrolabe, and having paid the modest price asked for it carried it home': ibid., 79.

45 Charles W.J. Withers, 'Mapping the Niger, 1798–1832: Trust, Testimony and 'Ocular Demonstration' in the Late Enlightenment', *Imago Mundi* 56 (2004): 170–93; Charles W.J. Withers, 'Geography, Enlightenment and the Book: Authorship and Audience in Mungo Park's African Texts', in Miles J. Ogborn and Charles W.J. Withers (eds) *Geography and the Book* (Farnham: Ashgate, 2010): 191–220.

46 NMS.H.MJ 205; See *Proceedings of the Society of Antiquaries of Scotland* 61 (1926–27): 262.

John Rae, the Canadian-based Orcadian sponsored by the Hudson's Bay Company, travelled to the Arctic to find out the fate of Sir John Franklin's expedition of 1848–9, and discovered their destiny in 1853–4. Franklin's men had died due to starvation, exposure and, perhaps, from lead poisoning from food stored in tins. Rae's octant is to be found in the Stromness Museum, in the Orkney Islands, together with many other important items relating to his work in Canada. National Museums of Scotland has a few items recovered by Rae, including Franklin's watch: most of these items are ethnographic and came to the Museum via the University of Edinburgh as part of that collections transfer in the late nineteenth century discussed above.[47]

James Bruce, explorer and linguist, is best known for his endeavours to discover the source of the Nile, reaching the headstream of the Blue Nile in 1770. He used the so-called Bruce Nile cup, a silver-mounted coconut shell cup, to drink the health of George III at Lake Tana, Ethiopia that year: presumably, he drank from the shell and had it suitably silver-mounted and engraved on his return.[48] But of the instruments he carried to explore Abyssinia, the sole survivor appears to be a camera obscura, and this remains with his family.[49] David Livingstone, perhaps Scotland's most famous missionary and explorer, will always be associated with his travels across Africa. Although none of his instruments is in the collection of National Museums Scotland, the David Livingstone Centre at Blantyre, near Glasgow, has a large amount of material relating to his life and times, including his microscope and a number of surveying instruments. These, together with material collected by Dr John Kirk, the economic botanist with him on his Zambezi

---

47   For items connected with Rae, see Ian T. Bunyan et al., *No Ordinary Journey: John Rae, Arctic Explorer 1813–1893* (Edinburgh and Montreal: National Museums of Scotland and McGill-Queen's University Press, 1993); also 'John Rae 200', in *New Orkney Antiquarian Journal* 7 (ed.) Sarah Jane Gibbon (forthcoming). For material collected by Rae and his contemporary, the geologist and naturalist, Dr James Hector, see Kaitlin McCormick, 'Early Collections of Haida Argillite Carving in Scottish Museums – A Preliminary Report', *Journal of Museums Ethnography* 27 (2014): 137–51.

48   NMS.K.2006.294. The provenance of this piece is as slender as it is for Robert Gordon's astrolabe. Acquired at auction, it had been held by the same Suffolk family for at least 80 years, and was probably purchased by them when the Fifth Lord Thurlow (head of the Cumming Bruce/Bruce of Kinnaird family) auctioned his family estates, including Kinnaird, in the late nineteenth century. His wife, Lady Elma Bruce, was a great-grand-daughter of James Bruce.

49   Bruce discusses the ordering of his camera obscura from Nairne & Blunt, London mathematical instrument makers, and how 'this, when finished, became a large and expensive instrument; but being separated into two pieces, the top and bottom, and folding compactly with hinges, was neither heavy, cumbersome nor inconvenient … its body was an hexagon of six-feet diameter, with a conical top; in this, as in a summer-house, the draughtsman sat unseen, and performed his drawings': James Bruce, *Travels to Discover the Source of the Nile, in the Years 1768, 1769, 1770, 1771, 1772 and 1773*, 5 volumes (Edinburgh and London: J. Ruthven and G.C.J. and J. Robinson, 1790): I, viii–x.

Expedition, 1858 to 1864, have been on display in Edinburgh several times during the past century.[50]

Conclusion

Changing perceptions of how collections are used often means that material that entered the museum in the past is not in what we might think of as the ideal form that is required by the curator of today for new displays. This raises questions to do with the tensions between the originating context of the object or instrument and the later interpretative uses to which it may be subject. It also raises questions about the curatorial responsibilities associated with an instrument apart from its display. At the point of accession, vital questions may not have been asked – over provenance, use, operational history, its association with significant moments or processes of exploration, or with notable explorers, even though the device itself may be commonplace. How far, then, is an instrument of exploration of long-term evidentiary significance – something worth having, worth displaying, worth telling stories about – merely by virtue of its associative connections, not for what, intrinsically, it is? How much do questions of manufacturing history or of its relative rarity make an instrument worth having? And when and how does the fact of having it translate into a requirement to display it? The curator imagines that explorers, in this case Scottish explorers, should be represented by the increasingly sophisticated scientific instruments that, in the context of their own time, enabled them to chart new regions – and is understandably surprised to find that such items are now hard to locate. In the case of National Museums Scotland this is the result of the institution's history, its collections policies, the emphasis afforded instruments as scientific objects relative to their ethnographic associations. In part also, it is a result, as for every museum, of too few funds being available to satisfy individual curatorial visions and institutional strategies. But we might also look to the past to explain present uncertainties in this regard. Is it the case that what we now interpret as devices that helped people explore, that extended human capacities in the measurement and ordering of nature, were in the past much less vaunted things? If this is so, could this be for the prosaic reason that what are now taken as 'instruments of exploration' were seen by (say) the late Enlightenment figures of Alexander Mackenzie, Mungo Park, and James Bruce as a means to an end, rather than as an end in themselves? Would Wyville Thomson, William Speirs Bruce, or Thomas Makdougall Brisbane expect the tools of their trade to

---

50   See, for instance, Alexander Galt, 'The Livingstone Centenary Loan Exhibition at the Royal Scottish Museum', *Scottish Geographical Magazine* 29 (1913): 243–52; *David Livingstone and the Victorian Encounter with Africa,* exhibition catalogue (London: National Portrait Gallery, 1996); Sarah Worden (ed.) *David Livingstone: Man, Myth and Legacy* (Edinburgh: NMS Enterprises, 2012). See also Lawrence Dritsas, *Zambesi: David Livingstone and Expeditionary Science in Africa* (London: I.B. Tauris, 2010).

be worth preserving? Is not the end product – the books, the papers in scientific journals – more significant than the means to such products' making? Attribution of worth which places 'value' – however that term is understood – more on the outcomes of the exploration and less upon the mechanics of its undertaking – risks keeping the instrument in storage more than on display. Such issues could explain why the type of exploratory material associated with key personalities that rests in collections is mostly of a natural history or ethnographic nature – and why the scientific instruments, which, in any case, arrived very late in the 'cabinet of curiosities' assembled in a particular Enlightenment tradition, remain without clear attribution of provenance, and, to a large extent, unrecognised.[51]

---

51   For the history of scientific instrument collecting, see A.J. Turner, 'From Mathematical Practice to the History of Science', *Journal of the History of Collections* 7 (1995): 135–50; R.G.W. Anderson, 'Connoisseurship, Pedagogy or Antiquarianism? What were Instruments Doing in the Nineteenth-Century National Collections in Great Britain?' *Journal of the History of Collections* 7 (1995): 211–25.

Chapter 7

'Instruments in the Hands of Others': The Life and Liveliness of Instruments of British Geographical Exploration, *c*.1860–*c*.1930

Eugene Rae, Catherine Souch and Charles W.J. Withers

This chapter presents the results of an instrument-centred analysis of the study of geography, technology, and exploration with reference to the role of the Royal Geographical Society (RGS) between the later nineteenth century and the first 30 years of the twentieth century.[1] The central importance of exploration to the emergence of geography as a science and to Britain's imperial interests in this period has been well documented.[2] Our central focus here is directed not at the

---

1   The phrase for the first part of our title, which signals both to the devices themselves and to the ways in which their human operatives were directed by others, is taken from a '*Review of Narratives of Travels and Discoveries in Northern and Central Africa*, by Denham Dixon, Hugh Clapperton and Walter Oudney', *The Edinburgh Review, or Critical Journal* 44 (June 1826): 173–219, quote from page 174.

2   For studies that stress the centrality of the RGS to geographical exploration in the British empire and of exploratory field science in the development of British geography, see Ian Cameron, *To The Farthest Ends of the Earth* (London: Jane and MacDonalds, 1980); David R. Stoddart, *On Geography and its History* (Oxford: Blackwell, 1987); Robert A. Stafford, 'Scientific Exploration and Empire', in *The Oxford History of the British Empire, Volume 3, The Nineteenth Century* (ed.) Andrew Porter (Oxford: Oxford University Press, 1999), 222–38; Felix Driver, *Geography Militant: Cultures of Exploration and Empire* (Oxford: Blackwell, 2001); Max Jones, 'Measuring the World: Exploration, Empire and the Reform of the Royal Geographical Society, *c*.1874–93', in *The Organisation of Knowledge in Victorian Britain* (ed.) Martin Daunton (Oxford: Oxford University Press, 2005), 313–36; Charles W.J. Withers, *Geography and Science in Britain, 1831–1939: A Study of the British Association for the Advancement of Science* (Manchester: Manchester University Press, 2010). For examples of British-led exploration and the role of the RGS, see D. Graham Burnett, *Masters of All They Surveyed: Exploration, Geography, and a British El Dorado* (Chicago: University of Chicago Press, 2000); Richard Drayton, *Nature's Government: Science, Imperial Britain, and the 'Improvement' of the World* (New Haven, CT: Yale University Press, 2000); Simon Naylor and James Ryan (eds) *New Spaces of Exploration: Geographies of Discovery in the Twentieth Century* (London: I.B. Tauris,

exploits of geographer-explorers or with the nature and timing of exploratory activity, but with the ways in which the RGS developed institutional strategies concerning the use of instruments. In particular, our intention is to understand how the RGS developed procedures towards the purchase and employment of precision instruments to be used in the field, how that body instructed and trained would-be explorers in the use of such equipment, and, especially, how the devices themselves fared in the course of being used. With respect to several instruments within the Society's collection, we document moments in the active working life of the devices and consider what the instruments' travels reveal about the nature of exploration. In order to do this, we have examined the Society's surviving manuscript minute books and committee records, paying particular attention to records of instrument purchase and remarks upon their use and to associated evidence concerning the development of instrumental instruction. We document the place of instruments and the development of guidance on their use in different editions of the Society's *Hints to Travellers*, first published in 1854. For the first time we scrutinise two unique manuscript sources in the keeping of the Society, namely the ledger volume 'Instruments Lent to Travellers', and the associated 'Catalogue of Instruments'. From assessment of these materials, it is possible to determine the RGS's institutional strategy towards instruments and to trace their movements over space and over time. In this way, we can discern how geographer-explorers and their precision devices together went to work in the world.

To consider instruments' institutional context and their travel and usage in these terms – and to focus less upon their users – is, we argue, to illuminate geographical exploration as something at once hard won and adventitious. In the first sense, exploration involving instrument use in the field demanded manual dexterity, robustness, and precision; at the very least, such things were presumed. It also encompassed issues of epistemological tolerance with respect to the results obtained. As others have shown, humans and instruments alike on occasion went slow or broke down, often in ways which cast doubt on their reported results, and so required repair to achieve effective operational standards.[3] In the second sense, the emergence of an instrumental culture within the RGS was itself a rather

2010); Dane Kennedy, *The Last Blank Spaces: Exploring Australia and Africa* (Cambridge, MA: Harvard University Press, 2013).

3   On such issues, see the essays in Marie-Noëlle Bourguet, Christian Licoppe and H. Otto Sibum (eds) *Instruments, Travel and Science: Itineraries of Precision from the Seventeenth to the Twentieth Century* (London: Routledge, 2002), and Simon Schaffer, 'Easily Cracked: Scientific Instruments in States of Disrepair', *Isis* 102 (2011): 706–17. On exploration and the physical and mental strains placed upon explorers, see Johannes Fabian, *Out of Our Minds: Reason and Madness in the Exploration of Central Africa* (Berkeley and Los Angeles: University of California Press, 2000). On illness and the conduct of science for one leading nineteenth-century figure, see Janet Browne, 'I Could Have Retched All Night: Charles Darwin and His Body', in *Science Incarnate: Historical Embodiments of Natural Knowledge* (ed.) Christopher Lawrence and Steven Shapin (Chicago: University of Chicago Press, 1998), 240–87.

halting process. Recognition of the importance of instruments to the advance of geographical science was not straightforwardly and at once paralleled by preparedness within the Society to buy instruments for explorers' use. Neither could instructions on use in the several editions of the Society's *Hints to Travellers* and in instructional classes be a guarantee of success or of accuracy in operation. Instrumental facts and figures purporting to exactness were often obtained in exacting circumstances. Instruments broke or were returned in poor operating conditions, then to be used by others. Given this, we may ask how reliable, instrumentally speaking, were the facts of geographical exploration when such exploration was often undertaken in difficult conditions, involved passage over terrain and not prolonged residence in it, and when it afforded little chance for considered reflection and corroboration until the explorer returned home.[4] Since these features were a commonplace of exploration, we need to recognise the complex agency of instruments when considering those processes that mediated the exploration itself and the production of explorers' published narratives.[5]

In elaborating upon these ideas, this chapter attempts to place the lives and liveliness of instruments centrally in our understanding of the nature of geographical exploration and in the workings of the RGS. To do so raises questions concerning instrumental usage, epistemology, and accuracy that lie at the heart of geographical exploration as a form of embodied practice, and raises questions, too, over the role of technology and instrumentation within geography's history. The chapter is in three parts. The first traces the role and significance of instruments, their purchase and their usage, within the operations of the RGS. The second examines the Society's rhetoric and regime of instrumental instruction, looking particularly at the work of John Coles, the RGS's map curator and instructor, between his appointment in the late 1870s and his death in 1910. The third analyses the 'Instruments Lent to Travellers' and the 'Catalogue of Instruments' to illustrate the use made of

---

4   Dorinda Outram, 'On Being Perseus: New Knowledge, Dislocation, and Enlightenment Exploration', in *Geography and Enlightenment* (eds) David N. Livingstone and Charles W.J. Withers (Chicago: University of Chicago Press, 1999), 281–94; Daniela Bleichmar, 'The Geography of Observation: Distance and Visibility in Eighteenth-Century Botanical Travel', in *Histories of Scientific Observation* (eds) Lorraine Daston and Elizabeth Lunbeck (Chicago: University of Chicago Press, 2011), 373–95. For a study of the conflicts between different routes to geographical knowledge and nineteenth-century British exploration of Africa, see Lawrence Dritsas, 'Expeditionary Science: Conflicts of Method in Mid-Nineteenth-Century Geographical Discovery', in *Geographies of Nineteenth-Century Science* (ed.) David N. Livingstone and Charles W.J. Withers (Chicago: University of Chicago Press, 2011), 255–77.

5   Ian S. MacLaren, 'Exploration/Travel Literature and the Evolution of the Author', *International Journal of Canadian Studies* 5 (1992): 39–68; Ian S. MacLaren, 'In Consideration of the Evolution of Explorers and Travellers into Authors: A Model', *Studies in Travel Writing* 15 (2011): 221–41; Charles W.J. Withers and Innes M. Keighren, 'Travels into Print: Authoring, Editing and Narratives of Travel and Exploration, *c.*1815–*c.*1857', *Transactions of the Institute of British Geographers* 36 (2011): 560–73.

different sorts of precision devices and to chart their geographical mobility and fractured lives. In conclusion, we address the implications arising from this focus upon instruments within accounts of geography's exploratory making, and of interpreting geography's scientific status in terms of technical accomplishment.

Exploration, Instruments, and the Emergence of Practice in the RGS before *c.*1860

At its foundation in 1830, the Royal Geographical Society established several committees to facilitate its work, including a Finance Committee, a House Committee, and a Committee for Expeditions: various ad hoc sub-committees, often of short duration, were also established. The committee minute books for the period 1830 to 1840, and from 1841 to 1865, have little of substance on the role of instruments and on their provision. An Instruments Committee was not established until 1859 (see below). Yet, the intention at foundation to obtain instruments for geographical work was clear: the third of the Society's stated initial aims records the need 'To procure specimens of such instruments as experience has shown to be most useful, and best adapted to the compendious stock of a traveller, by consulting which, he [sic] may make himself familiar with their use'.[6]

How soon the Society acted upon its stated aims, and what and whose was the basis in experience, is less clear. A report to the RGS Council in 1833 prepared by Julian Jackson, the Secretary, suggests that no instruments had been acquired by then despite initial intentions. It also highlighted differences within the Society over the steps necessary to promote the subject. Jackson's role in developing geographical methods and, particularly, guides to travel and advice on instruments for geographical exploration, has been examined elsewhere.[7] Jackson discussed whether the Society should focus upon publishing works on geographical topics including 'a manual for intending travellers', or should direct its energy into supporting exploration. He saw the latter as an opportunity to raise the profile of the Society, stimulate the interest of existing members, and encourage the recruitment of new members. Instruments were central to this exploratory emphasis, and, Jackson noted, they might be obtained by donation: 'A small collection of geodetical instruments sufficient to show the general nature of such operations, and of the means taken to accomplish them, besides being distinctly specified in our prospectus as a contemplated part of our establishment, would probably lead to a number of presents being made to us of new instruments as they progressively appear, and thereby, bring these under the notice of intending

---

6   [Anon.], *Journal of the Royal Geographical Society of London* (1831), 1, vi.

7   Driver, *Geography Militant*, 51, 53, 59; Charles W.J. Withers, 'Science, Scientific Instruments and Questions of Method in Nineteenth-Century British Geography', *Transactions of the Institute of British Geographers* 38 (2013): 167–79.

travellers'.[8] Jackson's hope that instrument makers would donate new instruments as they became available seems, however, to have gone unrealised. Certainly, Jackson was strongly of the view that the RGS had failed to deliver upon its initial objectives with respect to instruments: he noted how 'The 3rd point, the procuring of approved instruments, that travellers may be familiarised with their use has been completely neglected'.[9]

It is probable that the Society acquired instruments with a view to holding a working collection sometime after 1837, although there is no evidence to suggest that this was directly the result of Jackson's promptings.[10] Before 1859, dissatisfaction over the Society's instruments was formally registered through the Committee for Expeditions (sometimes also termed the Expeditions Committee) where concern over the availability and condition of instruments went hand-in-hand with anxieties over the Society's failure to provide travellers with adequate instruction in such instruments as were available. In the minutes of the Expeditions Committee for 18 November 1852, Francis Galton drew attention to what he simply called the 'want of proper instruments for travellers'. In the following meeting, on 20 December 1852, Captain Robert Fitzroy offered to draw up a list of proper instruments for travellers. Jackson had already done just this, in his 1841 *What to Observe; Or, The Traveller's Remembrancer.*[11] Fitzroy and Lieutenant Henry Raper were requested 'to draw up a set of general instructions for the use of travellers, to be laid before the Committee as early as convenient'.[12] The proper use of instruments was a key part of these instructions.

In these sources and for the period 1831–*c.*1860, we can discern two related themes. The first was a repeated anxiety over instruments and the almost complete failure to acquire them in any managed way. The second was the importance accorded textual instructions over instruments and their use as part of what an explorer should know. In the latter, we have the genesis of the RGS's *Hints to Travellers*, first published in 1854. As we show below, *Hints* can be examined

---

8   Royal Geographical Society (with Institute of British Geographers) Archives, MSS Additional Papers (hereafter, RGS-IBG Archives), AP 7, 2.

9   RGS-IBG Archives, MSS AP 8, 8.

10   There is evidence in its correspondence files that the Society bought instruments in its early years, but seemingly on an individual basis. In May 1839, for example, a Captain Symonds wrote to Captain Washington, RGS Secretary, to request the loan of instruments with respect to his (Symonds') planned expedition to New Zealand (which the Society had endorsed the previous year). Symonds asked for loan of a sextant, artificial horizon, boiling point and ordinary thermometers, and a pocket compass. Related sources record the purchase, in June 1839, of a sextant for Symonds' use at a cost of £9 9s: RGS-IBG Archives, Correspondence Block 2, Symonds to Washington, 3 May 1839 and 30 May 1839; Associated Papers Ledgers, June 1839.

11   Withers, 'Science, Scientific Instruments and Questions of Method in Nineteenth-Century British Geography', 171–3.

12   RGS-IBG Archives, Council Minutes 1841–1865, 56, 57, respectively.

to see which instruments were recommended, and why, and so document the Society's views over instruments and instrumental instruction.

The Society first established a committee to investigate the state of its instruments in 1859, probably in response to longer-run criticism over instrument provision, and in order to provide those items listed in *Hints* as necessary to geographical fieldwork. At the first meeting of the Instrument Committee on 18 February 1859, its initial agenda item concerned a circular to be sent to the main suppliers of the type of scientific instruments used in field research, such as Casella and Negretti and Zambra. Echoing Jackson's 1833 report, this circular invited suppliers to 'forward specimens of various meteorological instruments for inspection by the Committee', and additionally noted that 'barometer no. 36 in the possession of the Society was ordered to be sent to the makers to be tested at Kew preparatory to lending it to Mr Sewell'.[13] The fact that the barometer is numbered, with the implication that it was part of a catalogued set of RGS instruments, is, perhaps, of less interest than the evidence that devices were being sent elsewhere to be tested and calibrated and that the Society had adopted a loans policy for those instruments in its care. At its meeting of 25 February 1859, the Instrument Committee was tasked by the Society's Council to examine the existing instruments and produce a report upon them. This report, not included with the minutes (and which, unfortunately, has not survived), made clear the poor state of RGS instrument holdings: 'The Committee are clearly of opinion that the instruments in the possession of the Society are very defective and they recommend that a set of the meteorological instruments specified in the accompanying list should be purchased and kept as examples of what they recommend travellers to take'.[14] Even as the Society was identifying which instruments it wished to make available for the instruction of travellers in association with *Hints*, moves were made to dispose of instruments not up to scratch. Instructions over disposal first appear in the Finance Committee minutes for 19 March 1860, the reason given being lack of space within the Society's rooms.[15]

At the Instrument Committee meeting of 25 April 1860 a list of instruments, and, for some devices, a note of recommendation over their disposal, was included. There are 39 instruments listed, 27 of which bear non-consecutive identification numbers from a range 1 to 44 with a further 12 unnumbered instruments. It is probable that the instruments held by then, and those selected for disposal, effectively constituted a collection. This also suggests that, by about 1860, the Society had begun to take seriously the listing of its instruments and the need to keep records of their condition, and of the persons to whom, and for what

---

13   RGS-IBG Archives, Council Minutes 1841–1865, 157.

14   RGS-IBG Archives, Council Minutes 1841–1865, 160. The accompanying list includes maximum and minimum thermometers, wet and dry bulb hygrometers, Regnault's hygrometers, hypsometers, barometers and aneroids. Included with this minute is a small sketch showing a design for a thermometer carrying-case.

15   RGS-IBG Archives, Council Minutes 1841–1865, 191.

purpose, instruments were being loaned. From this date, we can be more certain over institutional developments. Between 1865 and 1910, there were only five years in which instruments were not purchased. Levels of expenditure increased over time, in relation (though not in direct proportion) to the levels of exploratory activity in which the Society's members were involved: in the 1860s, about £75 was spent on instruments; in the 1870s nearly £320; about £1060 in the 1880s; and about £2030 for the 1890s.[16] In August 1866, the Map Committee, responding to recommendations from the Finance Committee, set down the duties of the Society's map curator and for the first time formally acknowledged his role as keeper of the Society's instrument collection and his responsibility to accession new instruments and to keep a register of instruments lent.[17] In February 1870, the Finance Committee recommended to the Society's governing Council that members of the Expedition Committee, and not it, should have responsibility for selection of instruments. In December 1872, an Expedition Sub-Committee proposed limiting the types of instruments purchased by the Society to those recommended in *Hints to Travellers*, namely, sextants, artificial horizons, half-chronometers, compasses (both prismatic and ordinary), lanterns, thermometers (both ordinary and boiling point), aneroid barometers, and mapping instruments (hypsometers were added later to this list).[18]

Evidence on instrumental provision and RGS strategy before about 1860 thus allows us to make several observations. Early intimations over the importance of instruments to geography's development were not matched by their purchase, in any systematic and coordinated way at least, until about 1859. The establishment of the Instrument Committee that year was an attempt to coordinate, and, where necessary, replace the haphazard involvement of several different committees over instruments. A loans policy was not begun until 1860 although ad hoc lending occurred before then. For fuller understanding, we must turn to *Hints to Travellers* and other sources concerning the RGS's instructional culture with respect to instruments.

Geographers as Effective Instruments: Instructional Guidance and Instrumental Usage, 1854–1910

The RGS's *Hints to Travellers* drew upon an earlier publication, *Hints for Collecting Geographical Information*, which was reprinted with a preface by

---

16   These figures are taken from expenditure levels recorded in the Committee Minute books. The heightened levels of exploration recorded for the later 1870s and the 1880s and 1890s centred upon sub-Saharan Africa, mainly East Africa, Central Asia and parts of the Far East: see Jones, 'Measuring the World'.

17   RGS-IBG Archives, Council Minutes 1865–1873, 24 August 1866.

18   On this evidence, see *Hints to Travellers*. Fifth Edition (London: John Murray, 1883), 2–24.

Lieutenant Raper, additional papers by Captains Fitzroy and Smyth and 'Mr Baily's formula for calculating heights by the barometer'. The first use of the title *Hints to Travellers* appears in the minutes of the Expedition Committee for 7 February 1853 when Raper read his new preface, and the Committee recommended that the collected papers be sent for printing.[19] As Driver in particular has shown, *Hints to Travellers* must be interpreted within a wider context of instructive rhetoric in early science, including Jackson's 1841 *What to Observe*, and should itself be seen as an unstable attempt to impose authority over geography as an emergent, yet far from coherent, field of enquiry. As an instructive manual, *Hints* went through numerous editions from its first in 1854 in order to reflect different methodological concerns (such as, for example, the section on photography in the 1865 edition). Throughout its print life, however, *Hints* retained two features as a constant: 'an insistence on the need to record observations in a standardised form … and an emphasis on the use of reliable scientific instruments'.[20]

Here, we examine two principal features as they relate to one another and to the already examined RGS's strategy over instruments: John Coles' role as instructor, and his central involvement, together with the Hints to Travellers' Editorial Sub-Committee, in modifying the content and purpose of *Hints* over instruments.

Proposals to revise the third (1871) edition of *Hints* were first discussed by the Society's Council in January 1878. It was agreed that a memorandum be sent to numerous named persons and 'other principal travellers', asking for new material 'either in the way of arrangement, alteration, or addition; the main object being to supply as large an amount of useful information as possible, useful to travellers by land, in the simplest form and in the smallest compass, and especially such as is not to be found in ordinary tables of navigation'.[21] Among the 13 individuals identified and named was John Coles.[22]

Coles applied for the position of map curator in March 1877 (he had applied in 1872, but was then passed over in favour of Keith Johnson, the African explorer). His 1877 letter in application laid out his career in the Royal Navy since 1848 (which he had that year joined, aged 15). His record of service included work in West Africa, the south-eastern United States, the Pacific, the West Indies, the Baltic, and survey work on Vancouver Island. In 1872, he took up the position of nautical advisor to Hooper, the marine telegraph company, 'in which capacity I had everything connected either with Hydrography or Geography under my

---

19   RGS-IBG Archives, Council Minutes 1841–1865, 59.

20   Driver, 'Scientific Exploration and the Construction of Geographical Knowledge: *Hints to Travellers*', *Finisterra* 65 (1998): 21–30; *Geography Militant*, 56–67.

21   RGS-IBG Archives, Council Minutes 1877–1893, 3 January 1878, page 12.

22   The individuals identified in addition to Coles were: Rev. Thomas Wakefield; Rev. H.W. Holland; Mr Edw[ard] Whymper; Captain W. Gill; Major H. Trotter; Captain T.H. Holdich; Major General Baillie; Captain Woodthorpe; Mr Joseph Thomson; Mr W. Chandless, Mr John Ball; and Mr R. Spence Watson: RGS-IBG Archives, Council Minutes 1877–1893, 8 December 1880, page 188.

entire direction'. As he further reported, 'I understand theoretically and practically the projections used in all modern maps. I paint in watercolours. I have had considerable correspondence with foreign and colonial Governments. I have a practical knowledge of Nautical Astronomy, and should occasion require it I am able to express my ideas in public'.[23] This expertise underlies and explains Coles' central involvement in the development of instrumental instruction, in text and in practice, from the later 1870s.[24]

Coles' appointment must be considered in relation to other initiatives to develop the Society's professional expertise. Many stemmed from the presidency of Clements Markham. Markham established a Scientific Purposes Committee in June 1877. It was in this revitalised administrative context that the proposals to revise the then current edition of *Hints* were brought forward in January 1878. In June 1879, Markham submitted a further proposal for regulating the course of instruction offered to explorers and travellers by the map curator. This was limited to surveying and mapping, including fixing by astronomical observations, with pupils to pay 2s. 6d. per hour. Initial restrictions – that Coles as map curator-instructor was to be paid 5s. per hour and should not earn more than £15 per month for his instructional classes – were lifted in March 1881.[25] By October and November 1897, Coles was earning over £40 per month from classes in instrument instruction at a time when his monthly salary was less than half that amount.

Statistics on the numbers of persons instructed, by year, and by subject, do not survive as consistently maintained records. Yet we are afforded insight into Cole's instructional endeavours. In a letter to Galton in 1879, Coles tellingly reports how instruments fit for purpose were in short supply. Referring to his first pupil, a Mr Comber, Coles noted 'In the list of instruments, I see Mr Comber asks for two watches (half chronometers), with watertight cases; these I have not got, the

---

23   RGS-IBG Archives, Correspondence Block 6 1871–1880, 23 March 1877 [this mss is not foliated or paginated].

24   In a letter to H.W. Bates of 25 June 1879, Coles remarks that he was mulling over the conditions of his appointment [as Instructor] 'before definitely accepting the position'. But from evidence of correspondence with Bates in March 1879, it is clear that Coles had already begun to teach the use of instruments [to a Mr Comber] and that he [Coles] was to report to Bates upon Comber's expertise – 'that he [Comber] will be able to make proper use of the instruments, if furnished with them, by the Society': RGS-IBG Archives, Correspondence Block 6 1871–1880, John Coles, 11 March 1879. That Coles was giving instruction before he was officially established in the post is confirmed by a memorandum, 'Report of Instruction given, from April 16th to Nov. 12th 1880', wherein Coles remarks, with reference to Markham's memorandum of 1879, that since that time, 'no efforts have been wanting on my part, to give effect to the wishes of the Committee, and I trust to their satisfaction; I would therefore request (should the Committee see fit to do so), that my appointment as Instructor in Practical Astronomy and Surveying may be confirmed': RGS-IBG Archives, Correspondence Block 6 1871–1880, John Coles, 12 November 1880.

25   RGS-IBG Archives, Council Minutes 1877–1893, Scientific Purposes Committee, 27 June 1879, 21 March 1881.

Society have some watches and chronometers but none fit for the use of a traveller, they are old, their cases anything but tight, and all have seen much hard service, and now keep a very poor rate, under these circumstances I would ask you to authorize an order being given for the new watches, as those we have now would only mislead any one trusting to them'.[26] There was also a seeming diffidence among would-be explorers to be trained: 'it must not be lost sight of, that, many highly educated gentlemen are quite ignorant as to the very elements of Nautical Astronomy, and that many of them would, absolutely refuse to display their ignorance on the subject before any other than their instructor, and who would not be induced, under any circumstance, to join a class'.[27]

In July 1879, Coles was training Captain Temple Leighton Phipson-Wybrants in map work and in instrument use: 'Filling mercurial Barometers, and calculating Heights by Barometric differences, both by Baily's, and other Formulae. The practical use of the Hypsometrical apparatus and the calculation of heights by Boiling point Thermometers. Plain Trigonometry with reference to the measurement of heights and distances. Finding the Latitude and Longitude by two altitudes of the Sun, or by two altitudes of the same, or of different Stars'.[28] By late October 1879, Coles had three men under instruction: a Mr F.J. Rawson and a Mr Delmar Morgan in addition to Phipson-Wybrants. By mid-April 1880, Coles had trained a total of 10 men: 'four are members of the medical profession, one being the Government botanist in Afghanistan, the other three are missions [sic] to China and Africa, one is a Civil engineer in command of military expedition to the Upper Congo, one [Phipson-Wybrants] is an officer who has served with the rank of Captain in the army, and who is about to proceed to East Africa, another is a civil engineer who is preparing himself for foreign service by receiving instruction in the methods of fixing positions by astronomical observations, and my other pupil is about to re-visit Armenia and Persia in which countries he has already spent some time'.[29] Below, we identify the instruments that Phipson-Wybrants, Delmar Morgan, and several others took with them into the field and learn their fate – instruments and explorers alike.

Coles played a major role in the more-or-less constant revision of *Hints to Travellers* in the later nineteenth century. His first official involvement with revision work was for the 1883 fifth edition, on which matter, as noted, Coles was written to prior to his formal appointment. He recommended enlarging the

---

26   RGS-IBG Archives, Correspondence Block 6 1871–1880, John Coles to Francis Galton [undated, but, from internal evidence, April 1879].

27   RGS-IBG Archives, Correspondence Block 6 1871–1880, John Coles to H.W. Bates, 25 June 1879.

28   RGS-IBG Archives, Correspondence Block 6 1871–1880, John Coles, 31 July 1879, 1 September 1879. Phipson-Wybrants received 8 hours of instruction in August that year.

29   RGS-IBG Archives, Correspondence Block 6 1871–1880, John Coles, 'Report of Instruction Given, from January 22nd to April 16th 1880', 26 April 1880.

mathematical tables, and, in the absence of Captain T.H. Holdich (Holdich, a distinguished surveyor, later to be knighted and serve as President of the RGS, was then in India), Coles was given responsibility for the revision of the section on surveying.[30] His involvement increased with respect to the eighth (1901) and ninth (1906) editions, both as a contributing author, and, significantly, as the overall coordinating editor: Coles is given as the editor on the title page of both editions (having been listed only as sub-editor in July 1889 in a schedule of payments concerning the sixth edition, published that year).[31] His increased involvement with respect to the editions of 1901 and 1906 is to be explained by his own energies and expertise. As was the case in 1879 and the circumstances behind his appointment then, it can also be explained as the result of further moves from leading members of the Society to promote geography.

John Scott Keltie's 1886 report on geographical education and Halford Mackinder's 1887 paper on geography's cognitive reach were each significant moments in establishing a modern and university-based status for the subject in Britain.[32] This has been widely noted.[33] What has gone un-remarked upon is the renewed emphasis given to instrumental instruction as part of these initiatives. Specifically, it was recommended 'that the existing system of training travellers and observers be developed and completed'. This was a reference both to Coles' instructional teaching and to the improvement of *Hints*. As was noted, 'Our objects have been two fold – namely, to supply a course of instruction for observers in the field, and to place the geographical education of the country on a satisfactory basis. The first of these objects offers no difficulties; it is already partly secured, and the course can easily be made more complete'.[34] There are no surviving records of

---

30   RGS-IBG Archives, Council Minutes, 1877–1893, Hints to Travellers Editorial Sub-Committee, 8 December 1880, 10 February 1881, 23 November 1881, 20 June 1882.

31   RGS-IBG Archives, Council Minutes, 1877–1893, Hints to Travellers Editorial Sub-Committee, 2 July 1889. Coles was paid £50 for his revisions to this edition, in addition to his annual salary (£20. 16. 8d.), together with such remuneration as he received from instrumental instruction. In addition to his surveying revisions, Coles also contributed sections on boating and canoeing for the 1889 edition.

32   John Scott Keltie, *Report of the Proceedings of the Royal Geographical Society in Reference to the Improvement of Geographical Education* (London: John Murray, 1886); Halford John Mackinder, 'On the Scope and Methods of Geography', *Proceedings of the Royal Geographical Society* 9 (1887): 141–60.

33   David Stoddart, *On Geography and Its History, passim*; Charles W.J. Withers, 'A Partial Biography: The Formalization and Institutionalization of Geography in Britain since 1887', in *Geography: Discipline, Profession and Subject since 1870–An International Survey* (ed.) Gary Dunbar (Dordrecht: Kluwer Academic Publishers, 2001), 79–119; Ron Johnston, 'The Institutionalisation of Geography as an Academic Discipline', in *A Century of British Geography* (eds) Ron Johnston and Michael Williams (London: The British Academy, 2003), 45–90.

34   RGS-IBG Archives, Council Minutes 1891–1897, 14 November 1895, 'Scheme for a System of Geographical Instruction', 15 November 1895.

any changes made to Coles' instrumental instruction, but we can chart the changes made under his direction to the 1901 and 1906 editions of *Hints*.

Speaking in 1898 of the then current (and seventh, 1893) edition, Clements Markham noted that while it 'has undoubtedly been of great use and has been very highly appreciated', it was also 'rather too large for perfect convenience to a traveller, and contains much that might be condensed or left out'. Specifically, the brief outlined for the revision sub-committee – Coles, Cuthbert Peek (the astronomer and meteorologist who, in 1882, had endowed the Society with an award for geographers who made distinguished use of instruments in exploration), and Markham himself – was brevity with utility: 'to make a complete revision of the seventh edition, and to endeavour to reduce its bulk and cost without impairing its usefulness'.[35] Coles' report on the revisions made for the eighth edition reveals extensive changes and deletions, notably his own amendments and reductions to the surveying and astronomical portion, the 'Medical hints' entirely rewritten, the meteorology section rewritten, and so on.[36] In addition to producing *Hints* in two parts, each as a separate volume, and revising its content, the sub-committee under Coles' direction gave the go-ahead for embossed metric units of linear measurement to appear on the front cover and imperial units on the rear cover. In doing so, the Society's book of instruction was effectively made into an instrument (Figure 7.1). This utilitarian design featured on two further editions, in 1906 and on the tenth edition in 1921. In thus making the book itself a portable measuring device, and in guiding geographers to become effective instruments, Coles united the instructive rhetoric of Galton and Raper, to whom he was intellectual heir, with the practical, manual, and observational skills he taught to others, and the widely shared concerns of contemporaries for empirical enquiry through the trained use of precision devices. His obituary records that, in addition to his authorship for and editing of *Hints*, he instructed a total of 49 persons between 1879 and 1900.[37] As our final section demonstrates, it is possible to show for some of his 49 pupils and prescribed instruments what happened to them.

---

35   RGS-IBG Archives, Council Minutes 1897–1903, 1 February 1898.

36   RGS-IBG Archives, Council Minutes 1897–1903, 'Mr Coles' Report on the Present State of the Eighth Edition of Hints to Travellers'. This Report, a folio item, is separately affixed into the Minutes book, and is undated. From its positioning, it probably dates to the meeting of 13 March 1900.

37   E.A.R. [E.A. Reeves], 'Obituary. Mr. John Coles', *Geographical Journal* 36 (1910): 227–9. Coles retired through ill health in 1900, and died on 24 June 1910. His work on revising *Hints* was thus completed when, officially, he had ceased work for the Society just as it was begun before his formal appointment.

Figure 7.1   **The front cover of the eighth edition of *Hints to Travellers*
(London: John Murray, 1901) shows how the book, with
linear measurements embossed to the front (and rear) cover,
turned this manual of instruction into a portable manual
instrument of measurement**

Source: RGS-IBG Collections.

Instruments and Geographers at Work, 1861–*c.*1936

In order to illustrate the working lives of instruments held in the RGS's collections, and thus to document the fortunes of the devices and of their explorer operators, it is necessary to outline the two principal sources for this information.

The RGS Catalogue of Instruments is a handwritten record of scientific instruments purchased by the Society for lending to explorers whose expeditions were approved by the RGS.[38] Its first section, of 35 pages, contains a record of the instruments purchased, when, from whom, and for how much. The second section, pages 40 to 122, is a brief history of the use of the instruments. In between these two sections (on un-numbered pages) is a two-page index. Each entry in the first section contains a reference to a page number in the second section. Nearly 100 types and variations of types of instruments are listed in the first section. The instruments are arranged by type and range from direction/position/altitude-finding surveying devices such as compasses, sextants, artificial horizons, aneroid barometers and hypsometers to meteorological devices such as rain gauges and anemometers. The single craniometer in the collection was presented to the Society by one of its future presidents, Francis Rodd, on 8 February 1928, having been used by him in his two Saharan journeys in 1922 and 1927.[39]

The 'Instruments Lent to Travellers' volume itemises which would-be explorers borrowed which instruments and where in the world the intended exploration was to be undertaken.[40] Instrument loans numbering from 1 to 460 are listed, with a further 60 un-numbered loans recorded after the numbered items. The first recorded loan, in March 1861, is to Dr John Rae, the Orkney-born Arctic explorer, for his use in Canada and British Colombia: the last recorded loan is for July 1936, to Francis Rodd. Towards the rear of the ledger an un-numbered section – whose first entry dates from November 1883 – records the monetary value of the instruments lent against their value at their return. No reasons are given for the depreciation figures indicated and this section particularly is incomplete. Including the loans of two items for exhibitions and 14 for educational purposes, 571 instrument loans were made in the period March 1861 to July 1936.

It is unclear who had responsibility for these volumes and oversight of the entries made within them. Their date of commencement points to a close association

---

38　The volume, quarter bound in leather, measures 39 cm × 26 cm. There is no accession number for this item in the RGS-IBG Archives and it is here referred to simply as RGS CoI, with reference made to individual pages.

39　Francis Rodd, 'The Origin of the Tuareg', *Geographical Journal* 67 (1926): 27–47. As a practice within physical anthropology and surrogate instrument of racial 'science', craniometry was by then widely discredited.

40　This volume is the same dimensions as the Catalogue of Instruments, but has a different cover design. There is, likewise, no accession number and it is here referred to as RGS ILT, with reference made to individual pages, or, in relation to un-numbered pages, to the listed item or explorer concerned.

with the establishment of the Instruments Committee in 1859 and its work, in April 1860, to identify the instruments then held by the Society. There is uneven coverage with respect to the instruments' record (those little used generally having less related textual commentary). Cross reference between instruments and users is not always possible. The use of inks of different colours, and the presence of different handwriting, including that of John Coles and of his successor, E.A. Reeves, suggests that the ledgers are cumulative works of administrative record, some of it amended at later dates as the circumstances of explorer and instrument became known. Nevertheless, it is possible to know from these ledgers where in the world RGS instruments went and so to recover a relative historical geography of exploratory endeavour.[41] We can also identify who the explorers were, and what they took as instruments. For some instruments listed, it is possible to determine the geographies of their usage (Figure 7.2). And for a few explorers and instruments alike, we can trace lives and deaths. In these ways, we can begin to trace the lives of geographers as instruments of exploration and to trace the exploratory lives of instruments and to see how instructions given in one setting in text and by word of mouth, translated, effectively or not, in another.

Many of the more expensive precision devices purchased in the later 1870s were for use in African exploration as the Society turned its gaze to that part of the world.[42] In 1878, for example, a watch purchased from the London instrument makers Lund and Blockley at the cost of £35 was bought to support Keith Johnston's RGS-sponsored exploration of East and Central Africa. After Johnston's death in East Africa June 1879, it was used by Joseph Thomson.[43] Like Johnston, the watch never returned: Thomson left it in the keeping of Sir John Kirk in Zanzibar, who had travelled with Livingstone on the Zambesi Expedition of 1858–1863, and, in June 1899, the Society struck the item off its list of instruments.[44] The same fate was suffered by Coles' first pupil, the Rev. T.J. Comber, who died in West Africa in June 1887, and by the prismatic compass he used and which cost the RGS £5 5s: the instrument was struck off by order of Council in 1894.[45]

---

41   Excluding the 14 instruments lent for unspecified 'educational purposes', and the 2 for exhibitions, the regional breakdown is: Africa 229 instruments lent; Asia 127; Europe 54; Australasia and the Pacific 40; Arctic 36; South America 31; Antarctic 13; Central America 12; North America 12; Atlantic Ocean: RGS ILT, *passim*.

42   Jones, 'Measuring the World'.

43   James McCarthy, 'Alexander Keith Johnston (1844–1879)', in *Geographers Biobibliographical Studies* 26 (2007): 98–109. The fact that Johnston was absent from his RGS responsibilities as map curator, together with the delay in confirming the facts of his death in June 1879, explains why Coles was working as map curator and instrument instructor from late 1877 and why he wrote to RGS officers in November 1880 seeking formal confirmation of his appointment (see fn. 25 above).

44   RGS COI, page 2 (where the purchase is recorded) and page 43 (where the watch is struck off).

45   RGS COI, page 77. On the purchase of this prismatic compass, see page 14.

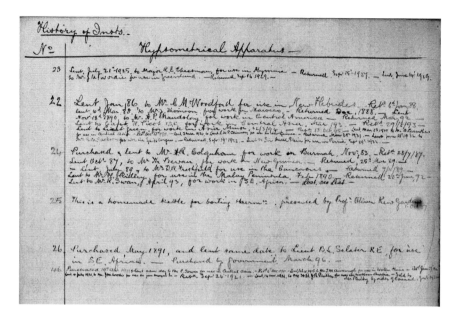

**Figure 7.2   This extract from the 'History of Instruments' section of
the RGS ledger 'Instruments Lent to Travellers' documents
moments in the 46-year trans-continental working life of
Hypsometer 22**

Source: RGS Instruments Lent to Travellers.

Fuller more successful instrumental and exploratory lives can be revealed.
Consider the example of Hypsometer 22 and its use recorded over 46 years. This
device, manufactured by the London firm of Cary, was first lent in January 1886 to
the ethnologist Charles Morris Woodford for use in the New Hebrides (Vanuatu).
He returned it in January 1888. In March 1888, it was lent to Joseph Thomson who
employed it in Morocco, returning it in December that year. In November 1890,
the geographer-archaeologist A.P. Maudsley (a member of the *Hints* revisions sub-
committee) took it with him to Central America, returning it in March 1892. Between
March 1893 and October 1895, this hypsometer was in Central Asia, being first
used there by Captain Picot, latterly by a Lieutenant Green. It was again in Central
Asia from March 1910, having been lent to Douglas Carruthers. Carruthers, by
then a veteran of expeditions in Africa and Central and Northern Arabia and holder
of the RGS Gill Memorial Award for his expeditionary work, was in many ways
the beau ideal of an RGS explorer, described by others as 'a young man who has
devoted nearly the whole of his time to scientific exploration'.[46] Carruthers took

    46   Douglas Carruthers, 'Exploration in North-West Mongolia and Dzungaria',
Geographical Journal 39 (1912): 521–51, qoute [by T.H. Holdich] on page 551.

the RGS course in instrument use and surveying in September 1909. As his plans developed for the exploration of Mongolia, the RGS Finance Committee in March 1910 approved the sum of £46 for the purchase of additional instruments.[47] Like many, Carruthers aimed to produce more accurate maps, correct extant accounts, and collect natural history specimens. He was, accordingly, dependent upon his instruments and attentive to their accuracy in the field and upon their return: 'The altitudes throughout the journey were taken with a hypsometrical apparatus (by Cary [this was RGS hypsometer 22], and in a few instances by Negretti and Zambra). In most cases readings were recorded from two thermometers, which were tested before leaving England and on my return, at the Kew Observatory. After 20 months' work in the field the thermometers read the same as they did at the start; and the same corrections have been applied throughout, when working the results'.[48] Hypsometer 22 was lent out again in March 1920, to Lieutenant E.W. Chinnery for surveying work in New Guinea. Chinnery retained it for 11 years, returning it in March 1931. In the following year, it was in Spitsbergen for four months, before, in September 1932, it was employed by Sir Mark Aurel Stein, best known for his work in Central Asia, for further study in Persia[49] (Figure 7.2).

Identifying individual instrument geographies in this way is a consequence of ex post facto scholarly reconstruction. It allows us to see that the utility of instruments in exploration was dependent upon their mobility and upon the training of operatives themselves as instruments. Contemporaries sought to trace the connections between exploration, instrumentation, and publication in the ways we have outlined. An RGS manuscript entitled 'List of Instruments Lent to Travellers 1877–1883' gives details of 30 individuals who undertook exploration in this period. Information is listed under 10 headings: the traveller's name; date [of exploration]; the place of exploration; instruments lent; value charged; value of instruments returned; results of observations published by the RGS.; results published elsewhere; results received but not published; and remarks. The originating authorship is not known, but in the 'Remarks' column, John Coles' handwriting, in darker ink, records the fate of some of the individuals listed, in places adding to the observations of the first author.[50] As for the two ledger volumes, this is a work of uneven cumulative record. We reproduce the upper portion of the manuscript here (Figure 7.3).

47   RGS-IBG Archives, Correspondence Block 7, Alexander Douglas Mitchell Carruthers, 26 September 1909, 9 March 1910.

48   Douglas Carruthers, 'Notes on the Maps Illustrating the Exploration in Mongolia and Dzungaria', *Geographical Journal* 41 (1913): 346–9, quote from page 349.

49   RGS ILT, pages 18, 87.

50   RGS-IBG Archives, AP 52, 'List of Instruments Lent to Travellers, 1877–1883'.

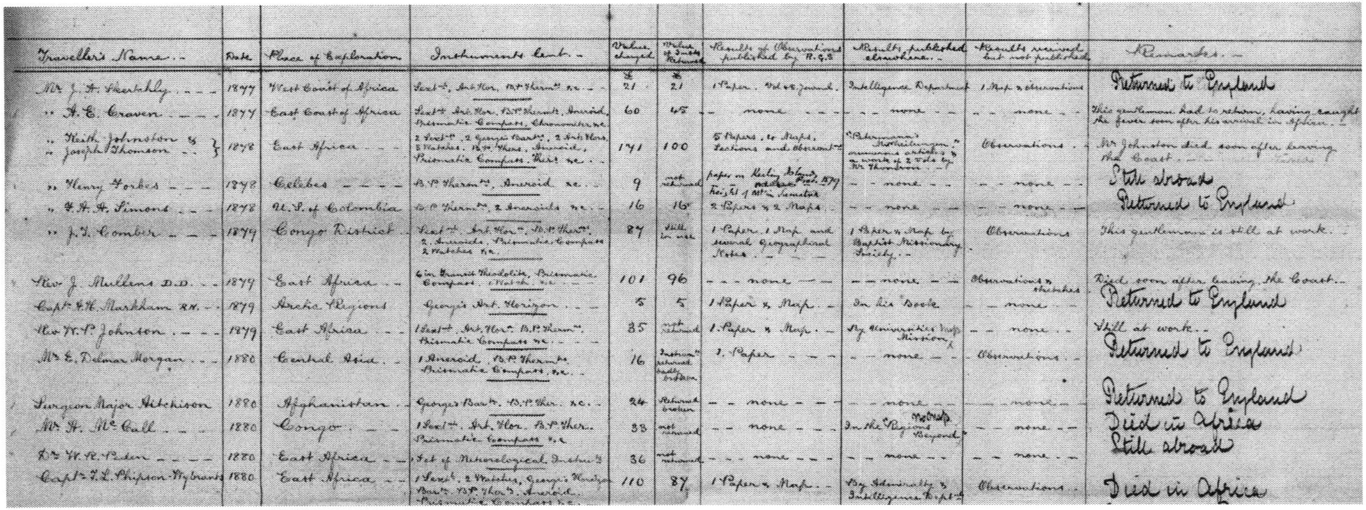

Figure 7.3 The life and death of instruments and of explorers is here illustrated in this detail from RGS manuscript 'List of Instruments Lent to Travellers 1877–1883'

Source: RGS-IBG Archives, AP 52.

The entry third from the top itemises what else the ill-fated Keith Johnston had with him in Africa other than the £35 watch (one of three). Trained as he was by Coles, and furnished with the equipment list as itemised in *Hints to Travellers*, Captain Temple Leighton Phipson-Wybrants was outlived by his instruments and his work. He died of fever in West Africa in February 1880; his listed contribution to the RGS (Figure 7.3, fourth column from the right) appeared in 1883.[51] Delmar Morgan, his one-time fellow pupil, returned his instruments 'badly broken'. Of the first 14 explorer-travellers listed, four did not return the instruments lent them; two others, Delmar Morgan being one, returned them in an unusable condition. Decades later, others found exploration to be similarly instrumentally taxing:

We had with us four other boiling-point thermometers, two of which had been lent us by the R.G.S. Only two were as a rule used, and those less frequently than we had intended, for fear of accident. Our best air temperature thermometer was broken on the first time of using it. This misfortune made us unduly nervous about our boiling-point thermometers, for it was essential to have them on the northern part of our journey. Our boiling-point stations were therefore limited in number, and for the immediate positions we trusted to two aneroids, lent us by the Society, and one by Casella which we found especially reliable.[52]

Coles' words to Galton in 1879 about the trust placed in instruments and the fragility of the devices had enduring resonance.

Conclusion: Instruments, Credibility, and Geographical Exploration

This chapter about instruments and exploration is an exploratory study in its own right. The manuscript sources examined, notably the ledger volumes, have not been the subject of study before and the potential they and the other sources examined reveal for a yet fuller understanding of the role of instruments in geographical exploration is obvious.[53] What does survive allows us to put the instruments that geographer-explorers used centre stage in understanding exploration as an

---

51   [Anon.], 'Obituary. Captain T.L. Phipson-Wybrants', *Proceedings of the Royal Geographical Society and Monthly Record of Geography* 3 (1881): 238–40; T.L. Phipson-Wybrants, 'The Delta and Lower Course of the Sabi River, According to the Survey of the Late Captain T.L. Phipson Wybrants', *Proceedings of the Royal Geographical Society and Monthly Record of Geography* 5 (1883): 271–4.

52   J.W. Gregory and C.J. Gregory, 'A Note on the Map Illustrating the Journey of the Percy Sladen Expedition, 1922, in North-Western Yunnan', *Geographical Journal* 62 (1923): 202–5, quote from pages 202–3.

53   We are delighted to record that Jane Wess is undertaking an AHRC Collaborative Doctoral Partnership PhD on the instrument collection of the RGS and its associated manuscript materials under our collective guidance, and we acknowledge Jane's helpful insights regarding our chapter.

accumulative process rather than an unproblematic accomplishment. Specifically, this study has established connections between institutional strategy, instrumental training and individual experience (for explorer and instrument), and, in a few circumstances, allowed us to trace the results of exploration in the form of publication, breakage, and death.

Several key points may be noted. The RGS saw instruments and their correct use as key features in the advance of geographical science. Their rhetoric in this regard, which dates from 1831, was not supported in a formal way until 1859 and the appointment of an Instrument Committee. Even the title of the instructive manual – *Hints* – suggests an early diffidence over the demands of exploration. Only from 1866 was the Society's map curator given managerial oversight of the instrument collection and a loan policy developed. The provision of instruments from the later 1850s onwards reflected wider educational initiatives within the Society, notably in 1859, 1879, and in 1895, the associated revisions of the Society's *Hints to Travellers*, and, from 1879, John Coles' role as instructor. Coles effectively turned *Hints* into an instrument of instruction and of measurement. In all likelihood, it was he who initiated the systems of record keeping with regard to instruments that are evident today in the surviving ledger volumes and the other sources here discussed. If Coles was an instrument in the hands of Markham and respective RGS presidents and others on Council, he also directed colleagues in the Society over its working practices on instruments.

These findings have wider implications. The study of nature through precision devices required training in their use, trust in their workings and readings, and faith in the results. It also depended upon others' trust in the truth of what one was told. For these reasons, exploration was something begun before one encountered the field, that had to tolerate breakdown and slow moving in the field, and something that was completed long after one had returned from the field (assuming one did): as was observed in 1937 by explorers returning their instruments to London, 'the completion of a survey takes about as long in the office as it has taken in the field'.[54]

To focus in the ways we have upon instruments and geographical exploration complicates that often utilised distinction in the historical geographies of science and of technology between 'the laboratory' and 'the field' as discrete spaces for the conduct of science.[55] There are further historiographical implications. Others have focused upon the 'hidden histories of exploration' and, in revisionist spirit concerning geography, exploration, and the RGS, have 'set out to highlight the role of local people and intermediaries, such as interpreters and guides, in making

54   Eric Shipton, Michael Spender and J.B. Auden, 'The Shaksgam Expedition', *Geographical Journal* 91 (1938): 313–36, quote from page 334.

55   For discussion of this topic and a review of relevant literature, see Charles W.J. Withers and David N. Livingstone, 'Thinking Geographically about Nineteenth-Century Science', in *Geographies of Nineteenth-Century Science* (eds) David N. Livingstone and Charles W.J. Withers (Chicago: University of Chicago Press, 2011), 1–19.

journeys of exploration possible'.[56] We echo and endorse these sentiments but here direct them at different intermediaries and at the travels of the instruments rather than always at those of their operators. Further and more detailed work might make it possible, of course, to see how indigenous human interlocutors and instruments as intermediaries worked together. As Coles and others knew in revising *Hints to Travellers*, precision devices were only one 'type' of instrument, and being trained in their correct use was only part of what was involved in readying one's self to undertake exploration. As men such as the Rev. T.J. Comber and Captain Phipson-Wybrants found to their cost, being trained was no guarantee of survival. It is precisely because exploration is and was precarious, involving passage over terrain, credibility and accuracy, the calibration of human and non-human agency, in-the-field error and the variable dissemination of results that has led others to question its utility as a way of knowing.[57] We have shown how its study in historical context with a focus on the lives of the instruments involved offers new insights into geography and into exploration.[58]

---

56   Felix Driver and Lowry Jones, *Hidden Histories of Exploration: Researching the RGS-IBG Collections* (London: Royal Holloway, University of London with the RGS-IBG, 2009), 5. On the topic of the human intermediary, see Simon Schaffer, Lissa Roberts, Kapil Raj and James Delbourgo (eds) *The Brokered World: Go-Betweens and Global Intelligence, 1770–1820* (Sagamore Beach: Science History Publications, 2009).

57   Vanessa Heggie, 'Why Isn't Exploration a Science?', *Isis* 105 (2014): 318–34.

58   We acknowledge the helpful comments of Sarah Millar and of Fraser MacDonald upon an earlier draft.

Chapter 8
Seismic Instruments, Geographical Perspectives

Deborah Jean Warner

The history of seismology – taking that term to mean the study, usually chronologically organized, of the systematic investigation of earthquakes – is relatively recent and well understood. An awareness of earthquakes and an interest in the Earth's interior were features of intellectual enquiry in Classical Greece and Rome, early China, the Enlightenment in Europe and what would become the United States. The establishment of the British Association for the Advancement of Science committee on seismology in 1842, and the foundation of the Seismological Society of Japan in 1880, have been held to mark the institutional expression of scientific interests into seismology as a branch of geophysics. But seismology did not become an independent science until the early twentieth century.[1]

To talk of the geographical study of seismographic instrumentation – taking that term to mean the study of the location of the manufacture, development, and testing of relevant instruments– is, by contrast, to propose a term so new as to risk any sense of what is understood by it. One reason for this is that in the histories of seismology, the devices used to measure earthquakes – that is, the technologies and instruments used to establish the subject's epistemological bases and scientific credentials – have been too often taken for granted. True, some histories of seismology note technical developments, but these developments are either lumped with the biographical assessment of foundational 'moments' and seismology's leading personnel, or they are accorded secondary status in interpretations that

---

1   On the history of seismology, see Charles Davison, *The Founders of Seismology* (New York: Arno Press, 1927); Markus Båth, *Introduction to Seismology* (Boston: Birkhauser Verlag, 1979); K.E. Bullen and Bruce A. Bolt, *Introduction to the Theory of Seismology* (Cambridge: Cambridge University Press, 1985); and David A. Valone, 'Seismology: Disciplinary History', in *Sciences of the Earth: An Encyclopedia of Events, People, and Phenomena.* Two volumes (ed.) Gregory A. Good (Garland: New York, 1998), 2: 745–9. As Valone notes, the emphasis to Davison's work is internalist, and the focus of Båth's account is on the epistemological development of seismology within the earth sciences more generally. Bullen and Bolt place the origin of 'modern' seismology to the mid- and late eighteenth century, prompted by the Lisbon earthquake of 1755 and the Calabrian earthquake of 1783.

assume histories of technological and epistemological developments as linear, progressive, and divorced from social or geographical context.[2] Another reason may be the uncertainty, itself historically conditioned, over what a scientific instrument is and, thus, what a seismological instrument is, and what have been the metrological and technical standards and languages behind seismology's emergence as a science. A yet further reason might be because it is only relatively recently that scholars have begun to take seriously the placed nature of science and of technology in historical context, to examine the social and technical spaces in which exploration has taken place, and in which the devices and instruments used in exploration have been put to work.

Within the last quarter of a century or so, a distinctive body of work has appeared that we can recognize as the 'geographies of science'. The focus of this work, to cite a leading philosopher of science, 'is on the conditioning of the practices and contents of the sciences by their national, institutional and geographical locations, and on the ways in which scientific knowledge has been distributed and mediated between sites of inquiry'.[3] In one review of work in the field, it has been proposed that, at its heart, lie

> two compelling sets of questions: the first having to do with the making and meaning of science *in place*, the second with science's movement *over space*. Why was science promoted *there* and not somewhere else? Why should science there have taken *that* form? How does science travel – within and between communities of practitioners, or from "expert" to "lay" audience? How is science communicated over distance?[4]

There is not the space here to examine the historiographical emergence of the spatial study of science in general, or these specific questions in detail. Neither should we suppose that one can uncritically replace the word 'science' by 'technology' and expect simply to arrive at something we might term the 'geographies of technology' of which the geographies of seismic instrumentation is a part.

Nevertheless, it is possible to propose geographical perspectives upon technology: to consider its placed and spaced nature; the settings – laboratories, workshops, factories – in which instruments were made and were made to do work; the ways in which technology travelled over space, to be received, modified,

---

2   On this point, see for example David J. Leveson, 'Whiggism and its Sources in Allègre's *The Behavior of the Earth*', *Earth Sciences History* 10 (1991), 29–37.

3   Nicholas Jardine, 'Chalk to Cheese: Progress, Power, Cooperation and Topography: Stages Towards Understanding How Science Happened', *Times Literary Supplement* (16 December 2011), 3–4, quote from page 3.

4   Charles W.J. Withers and David N. Livingstone, 'Thinking Geographically About Nineteenth-Century Science', in *Geographies of Nineteenth-Century Science* (eds) David N. Livingstone and Charles W.J. Withers (Chicago: University of Chicago Press, 2011), 1–19, quote from page 2.

pirated, improved, and so on. Examining technological issues spatially, here of devices whose purpose was to register, measure, and provide a material trace of sub-surface processes whose origin was not known, whose timing and duration could not be predicted, and whose consequences could be earth shattering, lets us see, more clearly than before, the situations in which seismic instruments were designed, produced, and deployed. It reminds us that these instruments were more scientific than mathematical; that is, they were made for and used by scientists who sought new ways of capturing the ever more elusive evidence of terrestrial tremors. This spatial perspective also shows that few of these instruments came from traditional instrument shops or centres of instrument production. Rather, most came from relatively small manufactories located near, and affiliated with, scientists who were particularly invested in this area of investigation. Early examples of seismic instruments originated in areas rocked by strong local tremors. Later, after scientists found that seismic waves pass through the earth, seismic observations became more dispersed, as did seismic instrument development and production.[5] The instruments in question include seismoscopes (which indicate that a seism or earth tremor has occurred), seismometers (which measure one or more aspects of seismic events), and seismographs (which record one or more measurements).[6]

Geographical perspectives on these issues also raise questions about disciplinary boundaries, or the lack thereof. Before seismology became an independent field of study, most practitioners saw themselves as physicists, geophysicists or engineers, and they tended to place their seismic instruments in astronomical or meteorological facilities. Further and related questions follow over identification, naming and dating, especially when seismic instruments were not so much invented as modified, redesigned and tinkered with to meet the critical challenges of the moment. And, since many instruments reflect complex collaborations, it is hazardous to attribute particular instruments, or even particular features, to any one person or, possibly, to any one specific site. We might better think of networks of collaborators working across space, sharing ideas and testing instrumental procedures, in order to arrive at commonly-held instrumental practices and working cultures.

This chapter explores the connections between the geographies and the technologies of seismic instrumentation and exploration. The first section briefly considers the local geographies of early seismic instrument making. The principal focus of what follows is upon national-level geographies of seismic investigation and on the relationships between academic institutions and the sites and practices of instrumental manufacture.

---

5   For pioneering work in this area, see Anna Lee Saxenian, *Region Advantage: Culture and Competition in Silicon Valley and Route 128* (Cambridge, MA: Harvard University Press, 1994).

6   James Dewey and Perry Byerly, 'The Early History of Seismometry (to 1900)', *Bulletin of the Seismological Society of America* 59 (1969), 183–227; Graziano Ferrari (ed.) *Two Hundred Years of Seismic Instruments in Italy (1731–1940)* (Bologna: Storia-Geofisica-Ambiente, 1992).

Early Seismic Instrumentation

Many early seismic instruments were made by or for the men who intended to use them. Following the series of earthquakes in and near Comrie in the southern highlands of Scotland in 1839, for example, the British Association for the Advancement of Science formed a committee to obtain 'instruments and registers' to record and measure these local shocks. This, in turn, led David Milne (later Milne-Home), a geologist educated at the University of Edinburgh, to coin the word 'seismometer', and James David Forbes, a natural philosopher at Edinburgh, to develop a vertical seismometer (so-called because it responded to the vertical component of terrestrial tremors).[7]

Robert Mallet, an Irish engineer and geophysicist, presented a paper 'On the Mechanics of Earthquakes' to the Royal Irish Academy in February 1846.[8] In a related paper, presented in June 1846, he expounded upon the value of 'certain new instruments' that he had devised 'for self-registration of the passage of earthquake shocks'. This, in Mallet's view, was crucial: 'Instruments previously intended for this purpose have not possessed the power of self-registration; they have consisted either in the trace left by the motion of a viscid fluid on the containing vessel, or they have been upon the principle of the inverted pendulum, or watchmaker's noddy. Instruments so constructed are objectionable, because having themselves times of vibration of their own, which may conflict with those of the earthquake shock, they are liable to fail in point of delicacy. They also possess several inconveniences of a mechanical kind in being adapted to self-registration'. For Mallet, the objects in view were to ascertain the time of transit of an earthquake shock; the altitude of the 'earthwave' at that moment; its horizontal amplitude; and its direction. His device involved a combination of mercury-filled tubes, galvanic batteries and an array of recording pencils: the length of the trace provided a 'graphic representation of the amount of the respective element of the wave' and a connected clock recorded 'the hour, minute, and second, and fraction of a second' at which the crest of the seismic tremor was registered, so allowing a measurement of the magnitude of the event and of its speed between two points of observation.[9]

Joseph Henry, the first notable physicist in the United States, used his position as the founding Secretary of the Smithsonian Institution to promote American science in general and seismology in particular. He had communicated with

---

7   James David Forbes, 'On the Theory and Construction of a Seismometer – An Instrument for Measuring Earthquake Shocks and other Concussions', *Proceedings of the Royal Society of Edinburgh* 1 (1845), 337–8; paper read on April 19, 1841. For the etymology, see Charles Davison, *Founders of Seismology*, 42–3.

8   Robert Mallet, 'On the Mechanics of Earthquakes', *Proceedings of the Royal Irish Academy* 3 (1844–1847), 186–92.

9   Robert Mallet, 'The Objects, Construction, and Use of Certain New Instruments for Self-Registration of Earthquake Shocks', *Proceedings of the Royal Irish Academy* 3 (1844–1847), 266–8, quote from pages 266 and 267.

Forbes for many years and kept abreast of his several investigations. It is thus not surprising that, when the US Congress authorized an exploring expedition to the southern hemisphere, Henry offered to purchase a seismometer made according to Forbes' designs and under Forbes' supervision, for use along the coast of Chile, an area that he saw as 'peculiarly favorable to the study of the facts connected with one of the most mysterious and interesting phenomena of terrestrial physics – namely the earthquake'. As it happened, this instrument did not perform as well as expected. Whether that failure was due to design, execution, or inexperience of the users is not clear.[10] Henry also had the Smithsonian reprint Robert Mallet's paper arguing that seismology had only recently become an exact science.[11]

Henry also had the Smithsonian publish instructions for making a simple seismoscope from a bowl filled with a viscous fluid such as mercury or molasses. He probably knew that Daniel Drake, a physician in Ohio, had employed a similar instrument to determine the direction of travel of the aftershocks following the New Madrid seismic event of December 16, 1811. In Drake's words, this was a device 'on the principle of that used in Naples, at the time of the memorable Calabrian earthquakes'.[12] De la Haute Feuille in France had described a similar instrument in the early eighteenth century, as did Cavalli in Italy in 1784.[13]

Alexander Dallas Bache, the second director of the US Coast Survey, was Joseph Henry's friend, and shared his enthusiasm for seismology. But while Henry had access to the Smithsonian's private funds, Bache was constrained

---

10   S.M. Gillis, 'An Account of Astronomical Observations Proposed to be Made in South America', *Abstracts of Papers Communicated to the Royal Society of London* 5 (1843–1850), 768–70, quote from page 770; S.M. Gillis [letter dated 24 October 1848], 'Letter from Lieut. Gillies [sic] U.S.N., to Lieut-Col. Sabine, R.A., For. Sec. R.S.', *Abstracts of Papers Communicated to the Royal Society of London* 5 (1843–1850), 990–4, quote from page 993; Joseph Henry to A.D. Bache, 18 January 1850, in Marc Rothenberg (ed.) *The Papers of Joseph Henry* (Washington: Smithsonian Institution Press, 1998), 8, 7–11; and Joseph Henry, 'United States Astronomical Expedition to the Southern Hemisphere', *Smithsonian Institution Annual Report* 8 (1854), 163–4.

11   Robert Mallet, 'On the Observation of Earthquake Phenomena', *Smithsonian Institution Annual Report* 13 (1859), 408–33.

12   Joseph Henry, 'Circular Relative to Earthquakes', *Smithsonian Institution Annual Report* 9 (1855), 245; Daniel Drake, *Natural and Statistical View; Or, Picture of Cincinnati and the Miami Country* (Cincinnati: Looker and Wallace, 1815), 234; Emmet F. Horine, *Daniel Drake (1785–1852): Pioneer Physician of the Midwest* (Philadelphia: University of Pennsylvania Press, 1961), 106. On Drake's value as a commentator, upon natural phenomena and more widely, see Michael L. Dorn, 'Volney's *Tableau*, Medical Geography and Books on the Frontier', in *Geographies of the Book* (eds) Miles Ogborn and Charles W.J. Withers (Farnham: Ashgate, 2010), 247–75.

13   For a study of contemporary reaction to, and representations of earthquakes where the instruments employed were the observers' (often disbelieving) eyes, see Susanne B. Keller, 'Sections and Views: Visual Representations in Eighteenth-Century Earthquake Studies', *British Journal for the History of Science* 31 (1998), 129–59.

by Congressional demands that the Survey's public funds be used for projects that would directly benefit ships in American waters. Bache was thus especially pleased when practical projects yielded scientific results. Such was the case with the self-recording tide gauges that the Survey installed along the California coast and that registered effects from an earthquake in Japan. Using these observations, Bache was able to determine the speed of the tsunami, the length of the wave at San Francisco, and the average depth of the ocean. And he established a tsunami warning program that has since expanded to cover the entire world.[14]

National Geographies of Seismic Exploration and Instrumentation

Japan was described by a local scientist described as 'preeminently a land of Earthquakes', and it became home to an important community of seismologists in the late nineteenth century.[15] A critical factor in explaining this was the Meiji government (the term means 'enlightened rule') that came into power in 1868 and that, seeking the benefits of modern science and technology, recruited a host of Western teachers and technical experts. After experiencing earthquakes, perhaps for the first time, some of these men developed seismic instruments, established observing stations, formed organizations, trained students, and laid the foundations for what would become the first sophisticated instrument enterprise in the Far East.[16]

This early seismological community in Japan was clearly international. One important member was John Milne, an English engineer whose interest in seismology had begun in 1876 when an earthquake roused him from his Tokyo bed. Four years later, after witnessing the devastation of Yokohama, Milne decided to dedicate his non-teaching time 'to elevating seismology from a geological pastime into a modern (instrument-based) science'.[17] To that end he spearheaded the establishment of the Japanese Seismological Society, and served as secretary of the British Association committee appointed for the purpose of investigating the earthquake phenomena of Japan. Working with Thomas Gray, a Scottish engineer teaching in Japan, Milne developed several seismographs. One recorded

---

14   A.D. Bache, 'Notes of a Discussion of Tidal Observations, Made in Connexion with the Coast Survey, at Cat Island, in the Gulf of Mexico', *Report of the Superintendent of the Coast Survey ... for the Year 1851*, Appendix No. 7, 127–36; 'United States Coast Survey', *North American Review* 90 (1860), 429–60, see http://www.tsunami.noaa.gov/.

15   Baron Dairoku Kikuchi, *Recent Seismological Investigations in Japan* (Tokyo: Dai Nippon Tosho Kabuishi, 1904), 1.

16   In the absence of a good history of the early Japanese instrument enterprise, one can look at educational instruments shown in Japanese pavilions at several Western international exhibitions in the late nineteenth century.

17   Gregory Clancy, *Earthquake Nation: The Cultural Politics of Japanese Seismicity, 1868–1930* (Berkeley: University of California Press, 2006); A. Herbert-Gustar and Patrick Nott, *John Milne, Father of Modern Seismology* (Tenderden: Norbury, 1980).

horizontal and vertical motions, first on a smoked glass and, later, in ink on paper. The Educational Article Manufacturing Company of Tokyo showed an example of an instrument of this sort at the Columbian Exhibition held in Chicago in 1893. Other examples were made by James White, an instrument maker who worked closely with William Thomson (Lord Kelvin), professor of natural philosophy at the University of Glasgow.[18]

National developments were, at least in part, the consequence of individual expertise in instrument design and manufacture as well as the dissemination and display of given devices. James Alfred Ewing, a physicist and mechanical engineer from Scotland, does not seem to have thought much about seismology before he was involved in the Meiji modernization of Japan. He soon found, however, that earthquakes occurred in Tokyo with 'a frequency which gives resident students of seismic phenomena an advantage difficult to overstate', adding that 'it is therefore no wonder that the instrumental side of Seismology has had its most considerable development in Japan'.[19] That is, he saw a direct connection between the geography of the phenomena and the development of instrumentation for its measurement. Writing to his mentor, William Thomson, Ewing boasted that his horizontal pendulum seismograph (it responded to the horizontal component of terrestrial tremors) was far superior to previous designs in 'simplicity and cheapness of construction, ease of use, and accuracy of results'. Ewing's companion vertical instrument was also a success. Ewing's '*Astatic* Horizontal Lever Seismograph' – designed to record tremors but not be itself shaken by them – was less successful, but did point the way to an important innovation.[20] By the mid-1880s, Japanese examples of Ewing's instruments were in regular use at the University of Tokyo and at the Japanese Meteorological Bureau. Later, after arranging for similar instruments to be made by the Cambridge Scientific Instrument Company, a firm that catered to scientists at the University of Cambridge, Ewing noted that

---

18 Thomas Gray, 'On Instruments for Measuring and Recording Earth-quake Motions', *Philosophical Magazine* 12 (1881), 199–212; John Milne and Thomas Gray, 'Earthquake Observations and Experiments in Japan', *Philosophical Magazine* 12 (1881), 356–62; Thomas Gray, 'On an Improved Form of Seismograph', *Philosophical Magazine* 23 (1887), 353–64; Japan, Ministry of Education, *Catalogue of Objects Exhibited at the World's Columbian Exposition Chicago, U.S.A. 1893* (Chicago: n.p., 1893), 55–9; T.N. Clarke, A.D. Morrison-Low, and A.D.C. Simpson, *Brass & Glass. Scientific Instrument Making Workshops in Scotland* (Edinburgh: National Museums of Scotland, 1989), 252–66.

19 R.T. Glazebrook, 'James Alfred Ewing, 1855–1935', *Obituary Notices of Fellows of the Royal Society* 1, 475–92. The quote comes from the preface to James Ewing, *Earthquake Measurement* (Tokyo: Tokio Daigaku, 1883). See also Neil Pedlar, 'James Alfred Ewing and His Circle of Pioneering Physicists in Meiji Japan', in *Britain and Japan: Biographical Portraits* (ed.) Ian Nish (Folkestone: Japan Library, 1999), III, chapter 8.

20 James A. Ewing, 'On a New Seismograph', *Proceedings of the Royal Society of London* 31 (1880–1881), 440–46, quote from page 443; James A. Ewing, 'On a New Seismograph', *Transactions of the Seismological Society of Japan* 1 (1880), 18–29.

these British instruments showed 'the high finish and perfection of workmanship' characteristic of the firm's work. And he thanked Horace Darwin, founder of the firm, 'for a number of suggestions the adoption of which has contributed much to scientific accuracy in details and simplicity in structural arrangements'.[21]

Thomas C. Mendenhall, an American physicist who believed that science should serve the public good, was introduced to seismology in Tokyo, and attended meetings there of the Japanese Seismological Society. After joining the instrument division of the US Signal Corps in the mid-1880s, Mendenhall collected earthquake information for the US Geological Survey. He used a seismometer of his own design to measure tremors caused by the breaking of an obstruction in the East River. And he put a seismoscope atop the Washington Monument to determine whether, and the extent to which this structure was affected by winds. Noting that 'Seismology as a science, or at least as an observational and experimental science, may be said to be the product of the past three or four decades', Mendenhall rejected the pioneering work of Mallet, turned to study single events not earthquakes as a whole, and saw in the United States advantages peculiar for their study: a generally intelligent population which provided 'a corps of willing and reliable observers', the telegraph network, and, most importantly, the 'almost universal use of standard time throughout the country'. He later brought his student, Charles F. Marvin, to the Signal Corps, and Marvin remained for many years with its successor organization, the US Weather Bureau. A skilled instrumentalist and able bureaucrat, Marvin designed and built a seismograph and operated it for several years. He also acquired two Bosch-Omori instruments for the Weather Bureau, and had that agency gather seismic information along with weather reports from around the country.[22]

After moving to Indiana as president of the Rose Polytechnic Institute, Mendenhall convinced William Ames, the professor of drawing there, to design a new seismoscope. By connecting this instrument to a clock, Mendenhall could determine the time of an earthquake that struck Indiana in 1887. He took an example of the instrument's reading to the National Academy of Sciences, and was

----

21   James A. Ewing, 'Earthquake-Recorders for Use in Observatories', *Nature* 34 (1886), 343–4; James A. Ewing, 'Seismometry in Japan', *Nature* 35 (1886–1887), 75–6; M.J.G. Cattermole and A.F. Wolfe, *Horace Darwin's Shop. A History of the Cambridge Scientific Instrument Company 1878–1968* (Bristol and Boston: Hilger, 1987), 34. Cambridge Scientific Instrument Co., *Physical and Electrical Instruments* (Cambridge: Cambridge University Press, 1902), 89–91.

22   Thomas C. Mendenhall, 'Seismoscopes and Seismological Investigations', *American Journal of Science* 35 (1888), 97–115, quotes from page 98 and 99. On Marvin, see Donald R. Whitnah, 'Charles Frederick Marvin', in *American National Biography* 14, 620–22; Charles F. Marvin, 'The Marvin Seismograph', *Monthly Weather Review* (July 1895), 250–22; Charles F. Marvin, 'The Marvin Seismograph', *Scientific American* (7 August 1897), 85; Charles F. Marvin, 'The Omori Seismograph at the Weather Bureau', *Monthly Weather Review* 31 (June 1903), 271–5.

no doubt pleased when the assembly endorsed the form.[23] And he hired Thomas Gray, whom he had met in Japan, to teach at the Rose Polytechnic Institute. Gray was no longer pursuing seismology, but did discuss a mantelpiece seismometer at the Indiana Academy of Science and at the American Association for the Advancement of Science.[24]

Italy is situated on another seismically active zone, and was home to another early community of seismologists. Luigi Palmieri, a professor of physics in Naples, designed an electromagnetic seismograph that recorded the time the event began as well as its duration, and was probably the first instrument to record tremors that were not otherwise perceptible. One example was placed in an observatory on the side of Vesuvius, another at the University of Naples, and yet another in Japan.[25] The Fratelli Brassart company, a Roman firm that specialized in meteorological instruments, began making electrical seismographs in the 1880s. One was devised by Achille Scateni, a mechanician in the physics department at the University of Urbino.[26] Luigi Fascinelli, head of the mechanical laboratory of the Central Office of Meteorology and Geophysics in Rome, would later make seismic instruments designed by Giovanni Agamennone, head of that organization.[27]

The first German seismological organization was formed in Karlsruhe in February 1880, after a moderate earthquake occurred in the nearby town of Kandel. This, in turn, led Ernst von Rebeur-Paschwitz, an astronomer at the University of Karlsruhe, to design highly-sensitive horizontal pendulum

---

23　Henry Crew, 'Thomas C. Mendenhall', *National Academy of Sciences, Biographical Memoir* 16 (1934), 33–5; A.A.A.S. Mendenhall, 'The Explosion at Flood Rock', *Science* 6 (1885), 326–8; A.A.A.S. Mendenhall, 'Observations and Opinions of the Savants and Points From Records: The Science of It', *New York Herald* (September 2, 1886), 3; Thomas C. Mendenhall, 'Report on the Charleston Earthquake', *Monthly Weather Review* 14 (August 1886), 233–5; Mendenhall, 'Seismoscopes and Seismological Investigations'; Thomas C. Mendenhall, 'The Seismograph', in 'Big Earthquake Wrecks Building in San Francisco', *New York American* (31 March 1898), 1.

24　Thomas Gray, 'A Mantle Piece Seismoscope', *Proceedings of the Indiana Academy of Sciences* 1 (1891), 20; Thomas Gray, 'The American Association for the Advancement of Science', *American Journal of Science* 40 (1890), 338; 'Thomas Gray', *Proceedings of the Indiana Academy of Sciences* 19 (1909), 24–5. See also John Milne, 'A Mantel-Piece Seismometer', *Transactions of the Seismological Society of Japan* 16 (1892), 47–8; and John Milne, 'Instruments for the Earthquake Laboratory at the Chicago Exposition', *Nature* 47 (1893), 356–7.

25　Luigi Palmieri, 'On the Electro-Magnetic Seismograph', *Smithsonian Institution Annual Report* 22 (1870), 425–8; Davison, *Founders of Seismology*, 89–90.

26　[Anon.], 'Electricity Applied to the Study of Seismic Movements', *Scientific American Supplement* 18 (1884), 7260–61; Fratelli Brassart, *Catalogo Descrittivo degli Istrumenti Sismici* (Rome: Leonardo, 1886); J. Wartnaby, *Seismology: A Brief Historical Survey and Catalogue of the Exhibits in the Science Museum* (London: HMSO, 1957).

27　George D. Lauterback, 'Giovanni Agamennone', *Bulletin of the Seismological Society of America* 38 (1948), 289–90.

seismographs, and Repsold, a noted instrument firm in Hamburg, to manufacture some seismographs of this type. Examining records from the example installed in the astronomical observatory in Potsdam, and seeing traces of what he believed was a Tokyo earthquake, Rebeur-Paschwitz made the important observation that seismic observations might be made anywhere in the world.[28] George Gerland, the professor of geography at the Kaiser Wilhelm University in Strasbourg, envisioned his institution as a centre for German seismology, and so arranged for a Rebeur-Paschwitz seismograph to be installed in the university's astronomical observatory. Rebeur-Paschwitz analyzed the seismograms from this instrument, concluded that earthquakes produced surface as well as body waves, and predicted, correctly, that the latter could be used to investigate the interior structure of the earth.[29] He also developed the proposal for an international system of earthquake stations that Gerland would present at the International Geographical Conference in London in 1895. This proposal led eventually to the International Association of Seismology, and to a Central Bureau of Seismology headquartered in Strasbourg. This development of internationalism in the institutionalization of seismological science was consistent with the development of international associations in other subjects and disciplines at this time.[30]

J. & A. Bosch, a small Strasbourg firm that produced meteorological and other measuring instruments, began making seismographs at about this time. One, the conic (or bifilar) pendulum instrument, was designed by Carl Mainka, manager of the instrument department of the Central Bureau of Seismology. Another was based on the design developed by Fusakichi Omori, a Japanese seismologist who had studied with Milne in Tokyo, and who maintained connections with seismologists around the world.[31]

International connections between national communities engaged in seismic instrumentation were developed in the German-Russian case through the figure of Boris Galitzin, a Russian prince who, while studying physics in Strasbourg, had come under Rebeur-Paschwitz's influence. Returning to St Petersburg in the mid-1890s, Galitzin became Director of the Physical Laboratory of the Imperial Academy of Sciences, established a seismographic vault at the Pulkova Observatory, and developed seismographs with two important features: electromagnetic damping and electromagnetic recording. This latter involved

---

28   Julien Fréchet and Luis Rivera, 'Horizontal Pendulum Development and the Legacy of Ernst von Rebeur-Paschwitz', *Journal of Seismology* 16 (2012), 315–43.

29   H.W. Duerbeck, 'The Observatory of the Kaiser Wilhelm University: The People Behind the Documents', in Andre Heck (ed.) *The Multinational History of Strasburg Astronomical Observatory* (Dordrecht: Kluwer, 2005), 89–122, quotes from page 109, 111.

30   Elisabeth Crawford, 'The Universe of International Science, 1880–1939', in *Solomon's House Revisited: The Organisation and Institutionalization of Science* (ed.) Tore Frängsmyr (Canton, MA: Science History Publications, 1990), 251–62.

31   J. & A. Bosch, *Seismische Apparate und Meteorologische Instrumente* (Strassburg: Klein, 1907); J. & A. Bosch, *Seismographen* (Strassburg: Klein, 1910).

attaching a coil to a seismograph in such a way that terrestrial tremors caused the coil to move through a magnetic field, thereby causing an electric current that would register on a galvanometer. With a mechanism of this sort, the recorder could be set up in a convenient place, while the instrument could be installed in a remote location.[32]

Hugo Masing, a mechanic affiliated with the Imperial Academy of Sciences in St Petersburg, developed and manufactured seismographs according to Galitzin's design.[33] Following Galitzin's death in 1916 and in the aftermath of the 1917 Russian Revolution, Masing moved to Tartu, in Estonia, established a Factory for Precision Mechanics, and served the needs of the local university. Johann Wilip, a physicist who had worked with Galitzin in St Petersburg, also moved to Tartu at this time and took up instrument design where his mentor had left off. Masing produced the first set of Galitzin-Wilip seismographs (one vertical and two horizontal) in 1925. In 1931, after completing a doctoral dissertation on 'Galvanometrically Registering Vertical Seismograph with Temperature Compensation', Wilip became the first Estonian professor of physics.[34]

The Geophysical Institute at the University of Göttingen, in Germany, was established in 1898 and led by Emil Wiechert, a physicist who had worked and published on elastic waves and on the size and mass of the earth, and who believed, as did Rebeur-Paschwitz, that seismic waves could be used to determine the structure of the interior of the earth. Wiechert also designed an inverted-pendulum seismograph.[35] Unveiled in 1900, this instrument was produced by Georg Bartels, a local firm that catered to the University, as had its predecessor firms for several decades.[36] In 1908, responding to a prize offered by the International Seismological Association, Spindler & Hoyer, another precision instrument firm in Göttingen,

32 Otto Klotz, 'Prince Boris Galitzin', *Bulletin of the Seismological Society of America* 7 (1917), 49–50; A.S. and G.W.W., 'Prince Boris Galitzin', *Proceedings of the Royal Society of London* 94 (1918): xxv–xxxi.

33 Harry Fielding Reid, 'On the Choice of a Seismograph', *Bulletin of the Seismological Society of America* 2 (1912), 8–30.

34 Johann Wilip, *On New Precision Seismographs* (Berlin: Schultz, 1926).

35 K.E. Bullen, 'Emil Wiechert', *Dictionary of Scientific Biography* 14, 327–8; Wilfred Schröder, 'Emil Wiechert and the Foundation of Geophysics', *Archives Internationales d'Histoire des Sciences* 38 (1988), 277–8; Joseph F. Mulligan, 'Emil Wiechert (1861–1928): Esteemed Seismologist, Forgotten Physicist', *American Journal of Physics* 69 (2001), 277–87; W. Schröder, 'Emil Wiechert und seine Bedeutung für Entwicklung der Geophysik', *Archive for the History of the Exact Sciences* 27 (1982), 369–89.

36 German Educational Exhibition, *Scientific Instruments* (Berlin: n.p., 1904), 13–14; Matteo Cerini and Graziano Ferrari, 'The Seismological Correspondence at the Observatory of the Alberoni College in Piacenza', *Annals of Geophysics* 52 (2009), 570.

introduced moderately inexpensive versions of Wiechert's seismographs suitable for smaller or less well-endowed institutions.[37]

While we may describe the geographies of seismic instrumentation in the later nineteenth century as 'national geographies', it is important to recognize the extent to which these national communities were internationally connected. This was, of course, because of the world-wide interest in the topic and the internationalization of science. And within these national pictures, certain universities and physical laboratories tended to dominate instrument and experimental work, often in association with local specialized manufacturing firms.

Sub-Surface Exploration and the Geography of Seismographic Instrumentation in the United States

Emil Wiechert had been interested in the practical implications and possibilities of seismology, but apparently 'needed one pupil to search for and find the way from theory to the practical, the profitable and with that the commercial application'. Ludger Mintrop, a mining engineer who began studying with Wiechert in 1907, was that man. After using a portable seismograph to locate the position of Allied artillery during World War I, Mintrop reversed the process, setting off explosions at a known distance from the seismograph. He applied for a patent concerning a seismic 'method of investigating the structure of strata' in 1919, and established Seismos, the world's first geophysical contracting company, in 1921.[38] The discovery in 1924 of the Orchard salt dome in Fort Bend County, Texas, by a Seismos crew prospecting for oil reserves prompted a race by others to steal that firm's secrets and to improve its instruments. Eventually, several similar firms were founded in and around the petroleum fields of the American southwest. The Geotechnical Corporation was one. Another was Geophysical Services Incorporated (which became Texas Instruments in 1951); this firm developed the technique known as reflection seismology in which sound waves set off at the surface could be used to identify sub-surface structures and, by interpretation, determine the likely presence of oil or gas reserves.[39]

---

37   Spindler and Hoyer, *Seismographen* (Göttingen, 1907); Spindler and Hoyer, *Communications Concerning a New Horizontal and Vertical Seismograph after Prof. Dr. Wiechert* (Gottingen, 1908).

38   Ludgar Mintrop, *Exploitation of Rock Strata and Mineral Deposits by the Seismic Method* (Hanover: Seismos, 1922); Ludgar Mintrop, *On the History of the Seismic Method for the Investigation of Underground Formations and Mineral Deposits* (Hanover: Seismos, 1930); Gerhard Keppner, 'Ludger Mintrop', *The Leading Edge* (September 1991); http://www.prakla-seismos.de/edge1.html and 'The Mintrop Mechanical Seismograph', *The Leading Edge* (September 1996), 1049–52.

39   George Elliott Sweet, *History of Geophysical Prospecting* (Los Angeles: Science Publications, 1966); J. Clarence Karcher, *The Reflection Seismograph: Its Invention and Use in the Discovery of Oil and Gas Fields* (n.p., 1974).

After a strong earthquake devastated parts of San Francisco in 1868, Joseph Henry had observed that the Pacific Coast will probably 'furnish an admirable locality for studying the phenomena of Earthquakes, and its savans, therefore, should give special attention to the subject'. A few years later, when the philanthropist James Lick was considering endowing an astronomical observatory in California, Henry recommended a facility that would support observations of a range of physical phenomena, including 'the registration of earthquake tremors'.[40] Following this advice, the Lick Observatory acquired a set of instruments from the Cambridge Scientific Instrument Co. in 1887. So too did the University of California at Berkeley. Edward Holden, who served simultaneously as Director of Lick and President of Berkeley, took particular note of the duplex pendulum seismometer that James Ewing had designed and which, Ewing had boasted, was 'employed by many private observers in Japan'. Hoping to see 'many such instruments distributed throughout the State', Holden convinced the California Electric Works in San Francisco to make copies for local distribution. Much as Mendenhall argued might be the case for the nation as a whole, this led to a network of seismic stations in the American West, each equipped with a $15 duplex pendulum, and each reporting their observations to the Lick Observatory.[41]

The massive earthquake that levelled much of San Francisco in 1906 provided further impetus for California seismology. In the words of one historian, this event tossed Harry Wood, a geology instructor at Berkeley, 'out of his bed and into seismology'. While working for the state commission charged with investigating the effects of the earthquake, Wood became convinced that a regional approach to seismic activity would produce the basis for local predictions.[42] Since commercial boosters hoping to attract settlers and businesses to their communities were reluctant to publicize the shakiness of the ground beneath their feet, local support was not forthcoming. But, from across the country, the Carnegie Institution of Washington was willing to provide funds so that Wood could establish a seismic programme at the Mount Wilson Solar Observatory, a Carnegie facility in southern California. It was here that John Anderson, an astronomer and instrument designer, designed a short-period seismograph that was reliable, compact and portable, and that was ideal for determining the location and depth of the epicentres of local earthquakes.[43]

40   Joseph Henry to Barton Stone Alexander, 27 February 1869, and Joseph Henry to James Lick, 13 December 1873, in Rothenberg, *Papers of Joseph Henry* 11, 228–31 and 482–5.

41   Edward S. Holden, *Handbook of the Lick Observatory* (San Francisco: Lick Observatory, 1888), 54–6.

42   Judith Goodstein, 'Waves in the Earth: Seismology Comes to Southern California', *Historical Studies in the Physical Sciences* 14 (1984), 201–30.

43   John A. Anderson, 'Seismometer', US Patent 1,552,186 (1925), assigned to C.I.W.; patent filed in 1924; Harry Wood and John A. Anderson, 'Description and Theory of the Torsion Seismometer', *Bulletin of the Seismological Society of America* 15 (1925), 1–72.

By 1930, 13 cities in the United States and one abroad had a Wood-Anderson seismograph of this sort. All were made by Fred C. Henson, a firm in Pasadena that employed no salesmen and produced no trade literature – yet which worked full-time to supply orders from all parts of the world for earthquake instruments.[44]

The success of this Mount Wilson program led the Carnegie Institution to endow a Seismological Laboratory. Established in 1927 and affiliated with the California Institute of Technology, this laboratory trained students, supported research, and served as the central station for a seismological network in southern California. It was here that Wood's assistant, Hugo Benioff, designed several important instruments.[45] Most were made by the Gilman Scientific Instrument Co., a small firm in Pasadena.[46]

Arnold Romberg was a graduate of the University of Texas who, after earning a Harvard PhD, worked as a seismologist in Hawaii. He returned to Texas in 1922 as an associate professor of physics and, like others on the faculty, supplemented his university salary by working as a consultant with petroleum firms. After the Humble Oil Company ordered two torsion balances from the Eötvös Collegium in Budapest, Romberg went to Hungary to learn how to use these instruments and to interpret their data.[47] In 1932, while teaching a graduate course in mechanics, Romberg asked his student, Lucien LaCoste, to design a long-period vertical seismometer. LaCoste came up with the notion of a 'zero-length spring' and, working with Romberg, built a crude instrument with this feature. In 1939, having realized that LaCoste's 'zero-length spring' could also be used in gravimeters (instruments widely used for prospecting purposes), LaCoste and Romberg went into the gravimeter manufacturing business. Since LaCoste had published an account of his central idea before seeking patent protection, other firms were free to copy his ideas – and they did. Almost all of the gravimeters made in the second half of the twentieth century, and there were many, were based on this key idea that came from seismology.

In the eastern United States, the Lamont Geological Laboratory was established in the late 1940s, affiliated with Columbia University, and located on a suburban estate just north of New York City. Here, with substantial funding from the Department of Defense, seismologists and others pursued scientific questions of particular interest to the Pentagon. Maurice Ewing was the founding director.

---

44   [Anon.], 'Dens of World's Earthquakes Searched For by Seismologists', *Christian Science Monitor* (15 December 1934), 4.

45   Hugo Benioff, 'Electrical Recording Seismograph', US Patent 1,784,415 (1930), assigned to C.I.W.; patent filed in 1927; Hugo Benioff, 'A New Vertical Seismograph', *Bulletin of the Seismological Society of America* 22 (1932), 17–31; Hugo Benioff, 'New Quake Gauge Success', *Los Angeles Times* (1 March 1932), A8.

46   Gilman price list in Frank Press papers, Box 4, folder 104, Massachusetts Institute of Technology Archives.

47   Arnold Romberg biography, see http://web.me.com/patandmel/UTexas_Physics_History/Arnold_Romberg.html.

His student, Frank Press, led the seismological programme. These men designed seismographs to capture the very long-period seismic waves that would provide information about the structure of the earth and, at the same time, record the distant detonation of atomic bombs. The first instruments employed to this end were made in-house. The Sprengnether Instrument Co., in St Louis, Missouri, made the production models. This firm had begun as a repair shop early in the century. Its involvement with seismology had begun in the 1930s, when William F. Sprengnether Junior enrolled at St Louis University, a Jesuit school that, like others of its ilk, had embraced seismology in an effort to obtain 'widespread publicity and general acclaim in a culture that celebrated science'.[48] The young student soon came under the influence of James Macelwane, an energetic and charismatic physicist who earned the first American PhD in seismology, from the University of California at Berkeley. More, perhaps, than any other single figure, Macelwane understood the benefit of a local manufacturing shop that could respond quickly and economically to specific research requirements.[49]

In the mid-1950s, Frank Press, director of the Seismological Laboratory at the California Institute of Technology, promoted a set of long-period seismographs that were remarkably similar to the Lamont instruments, but that were identified as 'Press-Ewing' instruments. Francis Lehner, the chief engineer of the Seismological Laboratory, contributed to the design, and his firm, Lehner & Griffith, made and marketed the instruments.[50]

In the 1960s, much seismic instrumentation and research in the United States was bound up in Vela Uniform (itself part of the larger Project Vela). This was a massive Cold War enterprise designed to detect underground bomb tests, and differentiate artificial explosions from natural seismic events.[51] The World Wide Standard Seismological Network, a key part of this project, consisted of 120 stations that were located around the world and equipped with identical apparatus. The long-period seismographs for this project were the Lamont form made by Sprengnether. The short-period seismographs were the Benioff form made by the Geotechnical Corporation, a start-up firm in Dallas, Texas, that would, in time, be bought by Teledyne, a major defence contractor.

---

48   Carl-Henry Geschwind, 'Embracing Science and Research: Early Twentieth-Century Jesuits and Seismology in the United States', *Isis* 89 (1998), 27–49; Perry Byerly and W.V. Stauder, 'James B. Macelwane, S.J.', *National Academy of Sciences Biographical Memoirs* (1958).

49   Deborah Jean Warner, 'The Sprengnethers and Their Seismographs', http://www.erittenhouse.org/artitcles/erittenhouse-vol-24-1-2012-2013/sprengnethers-and-their-seismographs/.

50   Deborah Jean Warner, 'Maurice Ewing, Frank Press, and the Long-Period Seismographs at Lamont and Caltech', *Earth Sciences History* 33 (2014), 333–45.

51   Kai-Henrik Barth, 'The Politics of Seismology: Nuclear Testing, Arms Control, and the Transformation of a Discipline', *Social Studies of Science* 33 (2003), 743–81.

The light-weight broadband seismometers that began to be used in large arrays in the later twentieth century in the United States and elsewhere follow the pattern laid out in this chapter. That is, they incorporated ideas proposed by academic scientists, and were made practical by instrument manufacturing firms that were relatively small, specialized, and often nearby. Thus, in central Europe, the so-called 'astatic leaf-spring instruments' were imagined by Erhard Wielandt, a professor of geophysics at the University of Stuttgart, and designed by G. Streckeisen of the Swiss instrument firm, AG Messgerate.[52] In the United States, the Ranger seismometers were based on ideas developed at the Seismology Laboratory of the California Institute of Technology for the ill-fated Ranger missions to the moon, and produced by Kinemetrics in California. Within a few decades of becoming a distinct science, seismology and its instruments had moved from the rather tremulous measurement of a trembling earth to devices which became a central part of hydrocarbon prospecting and, then, to be associated with the nuclear arms race and the space race.

Conclusion

Seismic exploration is a technical accomplishment. Understanding – even initially producing – the material traces of hidden large-scale physical phenomena requires instruments which, in Mallet's terms, 'self-register' the location, the horizontal and vertical dimensions of the event (its size), and its direction of movement. In this connected sense, the study of seismological instrumentation is an intrinsically geographical question: how and where did it become possible to measure the physics of the Earth?

This chapter has suggested that we may bring geographical perspectives to bear upon the study of seismological and seismometric instrument in several ways. At a local scale, this is a matter of understanding events which occurred in specific locations yet were recorded, in one way or another, in locations elsewhere: identifying the 'epicentre' of the earthquake event (a term coined by Mallet) from sites distant in space and time; or single recording sites in connected networks of seismographic laboratories. It is even a micro-local question inasmuch as the devices used to measure the earthquake event should not themselves be deflected or deranged in their working and recording of it by the event itself. While the specific design of instruments mattered here, so did the lay-out of seismological and physics laboratories and the positioning of instruments within them.

At the scale of the nation, it is possible to discern distinct geographies of seismological investigation – notably and initially in Japan, later in Italy, in Central Europe, and in the United States. These national geographies were never simply that. Many were internationally connected through the work of individual

---

52 E. Wielandt and G. Streckeisen, 'The Leaf-Spring Seismometer: Design and Performance', *Bulletin of the Seismological Society of America* 72 (1982), 2349–67.

scientists, by associational activities wherein the language and technologies of seismology was shaped, and by the commercial reach of manufacturing firms. Within national borders, seismometric investigation commonly took place in universities and laboratories working in association with nearby manufactories.

Thinking geographically about seismographical instrumentation as alone a matter of scale – of local, national, international, global, or extra-local geographies – may be productive in one sense. Yet it is also potentially reductive – geography as just a matter of site and place. It is clear even from this preliminary survey that the history of seismology and of seismographic instrumentation has been profoundly shaped by the movement of knowledge across space, by social interaction, by the collaborative integration of academic labour and industrial craft. The fact that seismic instrumentation has moved from the study of viscid liquids on table tops to bespoke laboratories to being part of Cold War geopolitics highlights a short but rich history as a distinct science and hints at geographies of scientific production and at distinct epistemic cultures within that science which would repay further and deeper exploration.

Chapter 9

Aerial Photography in Geography and Exploration

Peter Collier

The twentieth century saw the introduction of aerial photography as a tool for map-makers, geographers and explorers. For topographic mapping and some areas of thematic cartography, it would be reasonable to assert that the twentieth century was the era of aerial photography. However, the use of aerial photography for mapping and geographical analysis raised both methodological and technical problems, some of which have only been solved in the recent past. This chapter explores the conceptual, methodological, and technical issues around the use of aerial photography to make measurements (photogrammetry), the use of photography for analysis (air photo interpretation), and the role of aerial photography in the conceptual development of land systems analysis.

Aerial photography has been an attractive tool for map makers, geographers, and explorers for several reasons. An aerial photograph, or collection of aerial photography, allows an overview of an area, allowing spatial relationships to be established. Large areas of terrain can be covered easily, making mapping more efficient. The map-like nature of the vertical aerial photograph lends itself to both simple sketch mapping and to more precise mapping. It is more cost effective than land survey. Aerial photography provides a permanent record of an area at the moment of the photography, making it valuable in monitoring environmental change. Aerial photography can provide data for terrain that is inaccessible to on-the-ground exploration, or areas restricted for reasons of security or privacy.[1]

Although these are reasonable grounds to explain the utility of aerial photography, it was not until the late 1930s that aerial photography became widely used as a technology. A map of the state of world topographic mapping from the mid-1930s would not have looked so different to that from 1900. Most of Europe had been mapped, but that was not true of Africa, much of Asia, the Americas, or of Australia. The use of aerial photography, particularly after World War Two, completely changed this situation for reasons examined in what follows. By about 1970, most of the world had been mapped at a scale of 1:100,000 or larger. The development and use of aerial photography had a

---

1   Peter Collier, Dominic Fontana, and Alastair Pearson, 'GIS mapping of Langstone Harbour for an Integrated Ecological and Archaeological Study', *Cartographic Journal* 32 (1995): 137–42.

profound impact on the way people thought about the world, similar, perhaps, to the impact of the first pictures of the Earth taken from the Moon.[2] As few people had access to aerial photography in the early twentieth century, this impact was limited until after 1945. During World War II, propaganda or popular morale-boosting films, such as *Target for Tonight*, made by the Crown Film Unit, showed aerial photography being used for target identification.[3] In Britain, His Majesty's Stationary Office issued a series of propaganda publications, such as *Bomber Command*, *East of Malta West of Suez*, and *Coastal Command* that included reconnaissance photographs in their illustrations. After 1945, aerial photography was more widely used and better developed technically. Aerial photographs were published in newspapers and magazines, making the images available to the general public for the first time.

For aerial photography to become so widespread, to be publicly available, and, in time, to become a standard tool for geographers, either for map making or in terrain interpretation, several technical demands had to be met. This chapter outlines the development of these technologies of aerial photography in different national and institutional contexts. There everywhere needed to be a stable but flexible platform from which photographs could be taken. Cameras and films had to be suitable for use on those platforms. Appropriate measurement and interpretation techniques had to be devised to enable the extraction of the required information from the aerial photographs. These developments took place in relation to one another, although the problem of measurement was the last to be solved cost effectively. This is not, however, to present a teleological argument of technological progressivism or of a simple 'technical fix' solution to the problems of aerial photography and interpretation. It is to argue that the technical history and geography of aerial photography has been overlooked, and that one possible route to redress lies in understanding the technical challenges posed by taking pictures of the ground from the air. Where others have explored the association between technologies of surveillance and the earth's geographies as matters of 'ground truth', this essay outlines some of the technological issues to be overcome in securing 'air-borne truth'.[4]

2   Denis E. Cosgrove, 'Contested Global Visions: One-World, Whole-Earth, and the Apollo Space Photographs', *Annals of the Association of American Geographers* 84 (1994): 270–94.

3   For a discussion of this film, see K.R.M. Short, 'RAF Bomber Command's "Target for Tonight" (1941)', *Historical Journal of Film, Radio and Television* 17 (1997): 181–218.

4   John Pickles (ed.) *Ground Truth: The Social Implications of Geographic Information Systems* (New York: Guilford, 1994); Denis Wood, *The Power of Maps* (New York: Guilford, 1992).

Taking Photographs: The Need for a Platform

The idea of using photography for metrological purposes can be traced back to the work of Colonel Laussedat in France in the mid-nineteenth century.[5] His compatriot, Felix Tournachon (Nadar), who took the first aerial photograph in 1858, was also interested in developing the use of aerial photography for surveying and mapping. Tournachon was approached by the French Army which was interested in using balloons in its campaign in Italy, but Tournachon, a staunch republican, refused to lend his support to Napoleon III.[6] The technical limitations of the wet plate collodion process were anyway such that it was not a practical solution in aerial photography (a darkroom as well as a camera needed to be taken up in the balloon).

Although there were several civilian attempts to capture and use aerial photography in the last decades of the nineteenth century, most experimental work was carried out by the military. The role of the military in advancing the technology of aerial photography and its applications was to be of fundamental importance during World War I, World War II, and in the Cold War. Tethered balloons, free balloons, and kites were all used to varying degrees of success and most major armies deployed balloon or kite companies by about 1900. In Britain, Charles Close, a leading moderniser of survey practice as chief instructor at the School of Military Engineering in the early twentieth century and head of the Geographical Section of the General Staff before becoming Director General of the Ordnance Survey, had used balloon photography for mapping in India. Yet none of these techniques yielded significant numbers of useful images. Attention focused on the use of terrestrial photogrammetry with, in mainland Europe, significant work being carried out in the Alps by the Austrians and Italians, and, in Canada, by Deville.[7]

The development of the first aeroplanes had the potential to overcome the limitations of previous aerial platforms, but not until 1912 were aerial photographs taken explicitly for mapping. Captain Piazza of the Italian Army is usually credited with taking the first reconnaissance photograph in his photographs of the Turkish lines in the Turko-Italian War of 1911–1912.[8] Most major military-industrial powers were experimenting with aerial photography in the years before the First World War, but aerial photography was not used in that war until Lieut. George Pretyman of No. 3 Squadron Royal Flying Corps took aerial photographs over

---

5 A. Laussedat, 'Mémoire sur l'emploi de la Photographie dans le Levé des Plans', *Comptes Rendus* 50 (1860): 1127–34.

6 David S. Simonett, John E. Estes, and Robert N. Colwell (eds) *Manual of Remote Sensing* (Washington: American Society of Photogrammetry, 1983); Nigel Gosling, *Nadar* (London: Secker & Warburg, 1976).

7 Édouard Gaston Daniel Deville, *Photographic Surveying: Including the Elements of Descriptive Geometry and Perspective* (Washington: Government Printing Bureau, 1895).

8 Terrence J. Finnegan, *Shooting the Front: Allied Aerial Reconnaissance in the First World War* (Stroud: Spellmount, 2011), 22.

the Aisne battlefield on 15 September 1914.[9] With the stabilisation of the Western Front from late 1914, aerial photography became the only effective way in which intelligence could be gathered on activities behind enemy lines.

Following the increased use of aeroplanes for reconnaissance purposes, other aeroplanes were introduced to intercept them, and anti-aircraft guns were developed to shoot them down. The consequence of this was that aeroplanes flying with the intention of photography were forced to fly higher; this led, in turn, to the development of cameras with longer focal lengths. Increased aircraft speed also had an impact upon technical developments: photographs needed to be taken with faster shutter speeds to avoid blurring of the image. This meant that emulsions had to be faster. This latter problem was not finally overcome until the 1980s with the introduction of forward motion compensation systems in air survey cameras. In short, technical developments were simultaneously mechanical, aeronautical, and chemical, and all were initially driven by military imperatives.[10]

Initially, aerial photographs were taken with hand-held cameras, with the result being oblique photography. Following the realisation that vertical photography was more useful for measurement and mapping, it became common for cameras to be fitted to the fuselage of aircraft, or to cut a hole in the floor of the fuselage for the camera lens to look through. Once cameras were fixed to the fuselage, problems with vibration affected image quality, so vibration-absorbing mounts were developed. In these ways, the early technical requirements of aerial photography were dictated by military demands – for height, speed of flight, and clarity of image. Yet improvements in one area often led to problems in other ways.

Taking the Photographs: Developing the Films and the Cameras

To obtain useful images from airborne cameras, it was necessary to have a photographic medium which was less demanding than the wet plate process. The first practical medium utilised to this end was the dry plate process developed by Richard Leach Maddox in 1871, where the plate was coated with a photo-sensitive emulsion and left to dry. The plate could then be loaded into a camera and exposed at the required time. Processing the exposed emulsion could take place once the camera was returned to the ground. In 1889 the first photographic film was produced to replace glass negatives; glass continued to be used well into the twentieth century despite the obvious disadvantages.

By the early 1930s there had only been relatively minor improvements in the photographic process since the first introduction of the dry plate. Emulsions were chemically faster, that is, the exposure time was reduced. This was an important factor when taking photography from aircraft. A further major limitation on high

---

9   Finnegan, *Shooting the Front*, 43.

10   J.E. Farrow, 'Aerial Survey Camera Trials', *Photogrammetric Record* 12 (1986): 167–74.

altitude photography was the effect of atmospheric scattering on light. The back scatter of sunlight together with the light reflected from the surface of objects meant that, with the necessity of increased altitudes, the image provided by aerial photography was often reduced in contrast – that is, the ratio between darkest and lightest objects was reduced. In the ideal conditions of studio photography, the contrast, also called the useful object range, is normally about 1:300. In natural terrain the useful object range rarely exceeds 1:10. At around 7,000 metres scattering reduces the use object range to about 1:4, making accurate measurements and interpretation more difficult.[11] The solution adopted – to make the photography more sensitive at longer wavelengths and less sensitive at the shorter, blue, end of the spectrum where most atmospheric scattering occurred – was achieved by exposing the photographic emulsion to ammonia vapour. This extended sensitivity into the near infrared (the so-called photographic infrared between 0.7 and 0.9μm). This new emulsion was first tested on stratospheric balloon flights in North America and black and white infrared film was found to be particularly good at discriminating between different vegetation types, and between land and water surface. It was to be widely used for mapping high and low water lines around coastlines, and mapping land/water boundaries in areas covered by vegetation.

The next significant development was the introduction of Kodachrome, the first colour aerofilm capable of being used in air cameras. Kodachrome was a negative film, however, meaning that positive images needed to be printed from the film, making colour control difficult. The film was relatively slow and expensive, which slowed its adoption. From the perspective of interpretation, however, colour has significant advantages over black and white photography. The normal human eye can discriminate between about 40 grey scale levels between black and while, but between nearly 2.25 million different colours. Manufacturers and users were keen, therefore, to see more effective colour emulsions developed. The military in particular was aware that infrared film allowed discrimination between vegetation and camouflage netting which mimicked vegetation when imaged on black and white or normal colour photography. In 1942, Kodak developed Ektachrome, a reversal film (one where the film in the camera is processed through to a positive image). An infrared form of the film was quickly developed and introduced as camouflage detection film for military use.[12] There was considerable interest in the new colour infrared film after 1945, but cost remained a barrier to its widespread introduction. It did not come into general use until the late 1960s, but, from that period, it became the first choice film for environmental analysis until it was replaced by multi-spectral scanning in the later twentieth century.

---

11   Chief of the General Staff, *Military Engineering Volume XIII-Part X: Air Survey* (London: Ministry of Defence, 1979).

12   Kodak applied for the patent for its camouflage detection film on 24 October 1942, US Patent Number US2403722 A.

The basic idea of what an air camera should be was determined by the early 1920s. From that point, improvements were gradual. Experimental multi-lens cameras were designed to increase the area covered by single flight lines, but these were not widely adopted. More widely used was the trimetrogon system, which employed three cameras, one looking vertically, and two looking obliquely, one on either side of the aircraft. This system was widely employed by the US Army Air Force during the Second World War as part of its programme to create air navigation charts. After 1945, this system continued to be used in mapping programmes such as that used by the Directorate of Overseas Surveys mapping of Antarctica, until introduction of satellite imagery, such as Landsat, rendered it obsolete.

Paradoxically, one problem that arose with cameras in the inter-war period was that the demands of air forces for cameras for reconnaissance purposes conflicted with their use in mapping. Reconnaissance cameras need to have long focal lengths (up to 900 mm) and could have fairly small formats (that is, the negative could be only 120 × 120 mm) since the photographs were not required to give complete stereoscopic coverage. On the other hand, air survey cameras needed to give good overlap within the strip (typically 60 per cent overlap between adjacent photographs) and about 20 per cent between adjacent strips. In Britain, experimental work in the 1930s demonstrated that it was very difficult to obtain good stereoscopic cover using the types of cameras used by the Royal Air Force.[13] To ensure this stereoscopic cover, the format needed to be larger (typically 230 × 230 mm), and focal lengths shorter (152 to 305 mm). Cameras of this type came into use just before 1939 and were in general use by the war's end. To ensure that full stereoscopic coverage was obtained within the strips of air photographs, a degree of automation was introduced into the process of taking the photographs. Prior to the photographic sortie, the camera operator would calculate how far apart the photographs need to be taken to give 60 per cent overlaps knowing the required scale of the photography and the flying height of the aeroplane. The intended air speed of the aeroplane was then used to calculate the time between exposures, this being set on a device called an intervalometer. At the beginning of the strip, the operator would then switch on the intervalometer, which would then control the taking of the photographs and the advancing of the film in the camera.[14]

---

13   Peter Collier, 'The work of the British Government's Air Survey Committee and its Impact on Mapping in the Second World War', *Photogrammetric Record* 21 (2006): 100–109.

14   The required scale of the aerial photograph is related to the scale of mapping for which it is being flown. In general, the enlargement should be no more than four times between photographic and map scale; for example, aerial photography at 1:10,000 should not be used for mapping at scales larger than 1:2,500. The scale of aerial photography can be determined by the simple equation $S=f/H$, where S is the scale, f is the focal length of the camera and H is the flying height above the datum, usually sea level. The equation needs

With the introduction of 230 × 230 mm format cameras, the fact that the film did not lie flat in the focal plane led to distortion of the image. This was overcome by fitting a film-flattening device, initially a flat plate of photographic glass, which was pressed against the film prior to exposure. The action of this plate was also controlled by the intervalometer. Although it was initially found that when the film was wound on in the camera it generated static, which, if discharged, had the effect of lightening on the film, this problem was overcome by use of a vacuum device which sucked the film flat in the focal plane. Here, too, the story is of technical solutions creating problems that needed to be solved by other, often more complex and expensive, technical solutions. The result was that air survey cameras became increasingly expensive with the effect that fewer civil organisations or companies could enter the air survey market.

Using the Photography: Making Measurements and Maps

Developments in the technology for making measurements from photography encouraged attempts at map making for mountainous areas such as the Alps, where the Deutsche Östereich Alpine Verein started systematic mapping in the Austrian Alps. Similar work was undertaken in Italy where Pio Paganini in the Istituto Geografico Militare in Florence developed phototheodolites and simple instruments to assist in plotting from the photographs.[15] Scientists involved in Alpine mapping took advantage of new instrumental methods introduced by Carl Pulfrich in Germany and Henry Fourcade in South Africa.[16] The Pulfich instrument was called a stereocomparator. Fourcade called his a measuring stereoscope, but the name 'stereocomparator' was generally used of such instruments. Initially, these techniques relied on the measurement of individual points, the coordinates of which would then be calculated using parallax equations. While suitable for small areas and it was used by Fourcade to map Devil's Peak in Cape Town, it was not economical in mapping extensive areas. To make instrumental photogrammetry cost effective, automatic instruments were needed which would allow operators to trace line features and contours rather than measuring point by point. In Britain, Vivian Thompson's Stereo-plotter (1907), and, in Austria, Eduard Ritter von Orel's Stereoautograph (1908), did this from terrestrial photography. The von Orel instrument was adopted for alpine mapping almost immediately upon

---

to be modified in areas of high relief to $S=f/H-h$, where h is the mean ground height above sea level.

15 Teodor Blachut and Rudolf Burkhardt, *Historical Development of Photogrammetric Methods and Instruments* (Falls Church: American Society for Photogrammetry and Remote Sensing, 1988).

16 Peter Collier, 'The Impact on Topographic Mapping of Developments in Land and Air Survey: 1900–1939', *Cartography and Geographic Information Science* 29 (2002): 155–74.

its development. Thompson's instrument allowed Kenneth Mason of the Survey of India (later Professor of Geography in Oxford) to carry out photogrammetric mapping in the Pamirs in 1913.[17]

During the First World War, and after the Western Front had stabilised, aerial photography was an effective way to collect reconnaissance behind enemy lines or to map the disposition of military materiel. Mapping this information was not problematic as Northern France and Belgium were relatively well mapped. Relatively simple techniques could be used to undertake map revision using approximate methods, as long as common points were identified on existing maps and aerial photographs. Using such methods, the aerial photograph would be enlarged, either photographically or optically, to match the scale of the map to be revised. By matching between common points on the aerial photograph and the map, new detail could be added using perspective grids or superimposition of the photographic image on the map using epidiascopes.[18] If necessary, additional control points behind enemy lines could be fixed by conventional triangulation from behind one's own lines. The technical problem to be solved here, if the photographs were to be used for accurate mapping, was the precise determination of the camera station, that is, the position of the camera at the moment of exposure. Where common detail between existing maps and aerial photography was lacking, precise positioning was problematic. Although numerous methods were used to try to solve this problem, a complete solution was not found until the introduction of GPS in the early 1990s.[19]

Different geographies occasioned different technical solutions. Beyond Western Europe, where adequate mapping was limited, it was necessary to develop new techniques to determine the precise position over which the aerial photograph was taken. Two key theatres involving Britain were Egypt/Palestine and Mesopotamia. In Mesopotamia, mapping activities were largely the responsibility of personnel from the Survey of India. This is significant since, in the late nineteenth century and early twentieth century, the Survey of India was probably the most scientific and technically competent survey department in the world. This was due to the quality of the personnel recruited to work there. All senior positions were occupied by serving officers in the Royal Engineers, and these usually by those who had graduated top of their class from the School of Military Engineering. Locally-recruited staff included brilliant Indian mathematicians, who did most of the

---

17   Kenneth Mason, *The Thompson Stereo-plotter and its Use*, Survey of India Departmental Paper 5 (Dehra Dun: Survey of India, 1913); Kenneth Mason, 'The Stereographic Survey of the Shaksgam', *Geographical Journal* (1927) 70: 342–52.

18   Descriptions of these techniques are given in Malcolm Neynoe MacLeod, *Mapping from Air Photographs* (London: H.M. Stationery Office, 1920), and Lancelot N.F. Irving, *Graphical Methods of Plotting from Air Photographs* (London: H.M. Stationery Office, 1925).

19   Fritz Ackermann, 'Kinematic GPS Control for Photogrammetry', *Photogrammetric Record* 14 (1992): 261–76.

computational work. In order to create scaled maps from air photographs, it was necessary to locate positions of known coordinates in the area to be mapped. In the Near East, this was normally achieved by intersecting prominent points behind Turkish lines from known positions behind the British front line. In Mesopotamia, however, there were often too few suitable features to intersect. This difficulty was overcome by the use of an innovation by C.P. Gunter to fix control points within the Turkish lines using artillery shell bursts. Four observers would take simultaneous observations on individual shell bursts while an observer in an aircraft registered the position of the explosion on a plan of the trenches created from air photographs. On average, out of 10 shell-bursts, three good intersected points were observed and one of these would be marked accurately by the observer. Points observed in this way were then used to control and adjust the strips of photography.[20] In Palestine, the method used was to take a series of strips of photography parallel to the frontline and a number of control strips at right angles to the frontline using allied lines where surveyed control was available. This technique allowed the cartographers to create maps of comparable accuracy to the terrestrial plane tabling techniques usually employed.

Once the position of an aerial photograph had been determined, the next stage was to revise an existing map or compile a new one from the information on the photograph. It was quickly found, however, that existing plotting instruments were unsuitable for this work. Both the stereo-autograph and the stereo-plotter required that stereoscopic photography was co-planar, that is, that both photographs needed to be taken in the same vertical or horizontal plane. Taking coplanar terrestrial photography was relatively easy, but coplanar vertical photography was virtually impossible. This meant that plotting of detail was generally done graphically, by getting an approximate fit between the detail on the photograph and that on an existing map and then copying from the photograph to the map. But because photographs were inconsistent in scale, this made matching difficult. Inconsistency was due to two factors: the photograph not being truly vertical, which led to changes in scale across the surface of the photograph; and differences in ground height which meant that different points on the ground were at different distances from the camera, with those highest being at a larger scale than those which were lowest.

Geography also played a significant role in determining the suitability of different plotting techniques. On the plains of Northern France and Belgium, or in Mesopotamia, variations in ground height were usually insignificant. In Palestine it was a major problem. The main problem in the Western Front was tilt. As the aircraft used in World War I were light and so easily buffeted by the wind, cameras were hand held, at least initially. But this often led to tilted photography. Fixing

---

20 Peter Collier and Robert J. Inkpen, 'Mapping Palestine and Mesopotamia in the First World War', *Cartographic Journal* 38 (2001): 143–54; Peter Collier and Robert J. Inkpen, 'Photogrammetry in the Ordnance Survey from Close to Macleod', *Photogrammetric Record* 18 (2003): 224–43.

the camera to the fuselage reduced 'operator error', but could not reduce tilt due to the attitude of the aircraft. What was needed was some way of correcting for tilt. A solution, the process of rectification, had been proposed by Theodor Scheimpflug in 1903. Here, a tilted photograph is re-projected to remove tilt. But although the theory of rectification was understood, no simple rectifiers existed at the outbreak of war in 1914. The combatant powers were thus forced to use existing less precise devices, such as epidiascopes, or to change the scales of maps being copied. Efficient rectification was developed only after 1918 with the Zeiss SEG I in 1921, and from Odencrants in Sweden in the 1920s. These technical developments drove requirements enhanced levels of training and skill for those using aerial photography. Yet this trend was, in turn, more marked in Western and Central Europe than it was in the English-speaking world, where reliance continued to be placed on low technology solutions. This trend was itself only reversed in the 1980s with the introduction of computer-assisted methods.

In the early twentieth-century United States, the comparative lack of quality up-to-date mapping made the use of aerial photography an attractive proposition. The United States Geological Survey and the Coast and Geodetic Survey had made use of aerial photography as early as the 1920s. James Bagley created maps using radial line plotting techniques, although these were not yet on a theoretically sound basis.[21] The availability of redundant military aerial cameras and aeroplanes meant that a number of air survey companies were established after 1918. Coastal wetland and inland marshlands were among the first environments to which these new techniques were first applied.[22] In Austria, to offer one contrast, the break-up of the Habsburg Empire and the disbanding of the Militärgeographische Institut in Vienna were severe setbacks to the development of air survey, but scientific institutions within the former Habsburg lands of Poland, Hungary and Czechoslovakia gained trained and experienced cadres who were able to form the nuclei of the new survey departments.[23] In the Dutch East Indies, the use of aerial photography was quickly adopted as a way to map vast previously unmapped areas, despite the Netherlands being slow to realise its potential. The technology of aerial photography had distinctly different geographies and chronologies of up-take.

In Britain, despite having trained and experienced personnel, there was, initially, little official enthusiasm for air survey. The employment of O.G.S. Crawford by Charles Close was important for the development of aerial archaeology in

---

21   Committee on Photographic Surveying of the Board of Surveys and Maps of the Federal Government, *The Use of Aerial Photographs in Topographic Mapping: Air Service Information Circular 2* (London: Government Printing Office, 1921).

22   L.D. Graham, *Topographic, Hydrographic and Aerial Survey of Lake Okeechobee, Florida. C&GS Annual and Season Report 30*, 1924–25; G.C. Mattison, *Aerial Survey of the Mississippi Delta* (London: Government Printing Office, 1924).

23   Ingrid Kretschmer, 'The Mapping of Austria in the Twentieth Century', *Imago Mundi* 43 (1991): 9–20.

Britain, but this more specialist work was not valued by all.[24] Despite official lack of interest, numerous individuals pressed the case for the more systematic development of air survey. One such advocate was Stewart Newcombe, a Royal Engineer officer who had been involved in terrestrial survey in Sinai and Palestine before the First World War. He was an intelligence officer during the war, and worked with T.E. Lawrence on his missions with the Arabs. After 1918, Newcombe took command of the 7th Field Survey Company in Palestine and there quickly recognised the potential of air survey for geographical and political surveillance. Later, as Chief Engineer in Malta, he was to organise the first systematic air survey. At the prompting of Newcombe and others, the decision was made to establish an Air Survey Committee to research the potential of aerial photography and survey, primarily in a military context. Little regard was given at the outset to its potential civil applications. The research officer appointed in 1925 to carry out experiments and to design equipment was Martin Hotine, arguably the most gifted British surveyor of the twentieth century. His appointment led directly to the development of radial line control and plotting techniques on a sound theoretical basis.[25] This underpinned much of the medium scale mapping work carried out in the English-speaking world for the next 50 years. The chief advantage of Hotine's radial line technique was that it used very simple equipment, little more than a straight edge and a sharp pencil. This was technology at a basic but effective level. The technique was easily taught and, although labour intensive, lent itself to the mass production of mapping using relatively unskilled staff. The 1:50,000 topographic mapping of much of Anglophone Africa between the 1940s and the 1960s was mostly carried out using this technique.

In mainland Europe by contrast, the main focus was on the development of instrumental photogrammetry. Two basic approaches were adopted; direct optical projection and mechanical restitution. In the former, a stereoscopic pair of photographs is projected using specially designed projectors. The two projected images were manipulated to reconstruct the spatial relationships that exist between the ground and the two camera stations from which the images were taken. When this had been accomplished, the photogrammetrist could view a theoretically correct model of the ground, and so plot detail and contours. This concept had been proposed by Porro and Koppe in the nineteenth century in conjunction with terrestrial application, but was found difficult to implement in the context of aerial photography. The first practical solution was proposed by Max Gasser, with a prototype being manufactured in 1917. A fully practical solution was finally

---

24   On the work of O.G.S. Crawford in the development of aerial photography in archaeology, see Kitty Hauser, *Shadow Sites: Photography, Archaeology and the British Landscape* (Oxford: Oxford University Press, 2007); Kitty Hauser, *Bloody Old Britain: O.G.S. Crawford and the Archaeology of Modern Life* (London: Granta Books, 2008).

25   Martin Hotine, *Simple Methods of Surveying from Air Photographs* (London: H.M. Stationery Office, 1927).

developed through the use of the goniometer, projectors that could be rotated around three orthogonal axes.

Even once a practical instrument had been developed, what was still needed was an easy way to orient the photographs to create the stereoscopic model. Because aerial photographs were not taken truly vertically, any orientation procedure needed to recreate the tilts at the moment of exposure, and these tilts would be different at each camera station. The lack of correspondence between the two projected images resulted in parallax in the resultant model, leading to inaccurate measurements.[26] There were two types of parallax; X parallax in the direction of flight, which could be removed by altering the separation of the two projectors, and Y parallax, at right angles to the X direction, which prevented an operator seeing a model. Once the parallax had been removed, it was necessary to orient and scale the model relative to ground control. For this the operator needed a minimum of three known plan points and four known height points. The coordinates of these points were provided by ground survey. Although the work of Otto von Gruber and others in the 1920s meant that a theoretically correct model of the surface could be created quickly, so making photogrammetry more cost effective, it was still dependent on expensive ground control to achieve an accurate solution. Only with the advent of global positioning satellite systems in the 1990s was the problem of providing cheap ground control overcome.

The alternative to the direct projection approach to model formation was a mechanical solution. In instruments of this kind the projector lenses and the rays of light they project were replaced by a mechanical analogue of the lens, a gimbel or carden joint, and the rays of light by space rods which pass through the gimbels. In general, mechanical instruments provided much better viewing for the operator, but required frequent calibration to ensure the optical and mechanical components were correctly aligned. After 1945 most important civilian developments in photogrammetry took place in Europe, with Zeiss (Oberkochen) in West Germany, Zeiss (Jena) in East Germany, Wild and Kern in Switzerland, Galileo in Italy, and Poivilliers in France being the most important firms. Instruments mainly based on either directed optical projection or mechanical projection dominated instrumental photogrammetry until the widespread introduction in the 1980s of analytical instruments which used computers to assist in the orientation and measurement processes. Instrumental photogrammetry required a much higher level of training and skill in the operator than that required by the radial line method. Photogrammetrists in the US Geological Survey were required to graduate in civil engineering, and, in Britain, those in the Ordnance Survey were, generally, experienced land surveyors once it adopted instrument photogrammetry, although, thanks to Hotine's elegant technical solutions, radial line methods could be quickly learned by anyone with a steady hand and good stereoscopic vision.

---

26 Parallax is the apparent differences in position of two objects at difference distances from the observer when seen from two different positions. It is how we perceive objects as being at different distances from ourselves.

The problem of ground control was partially solved with the introduction of the Zeiss multiplex instrument (Figure 9.1). The multiplex had a number of projectors which could be moved along a bar and which could be oriented to create stereoscopic models. In addition to the orientations around the three axes, each projector could also be translated in X, Y and Z dimensions. This meant that after an initial model had been formed and adjusted to fit a platform of ground control, a third projector could be oriented to fit the existing model using a technique called single projector orientation. Additional models could be tied on sequentially using single projector orientations until the end of the strip was reached. Typically, up to 13 projectors could be used, creating 12 models. But, as the human operatives proceeded along the strip, more and more errors could be generated in the models as the strip would gradually lose orientation in its X and Y dimensions, and, in particular, in the Z dimension. To overcome this, a second platform of control was needed for the last model in the strip, and additional height points were needed

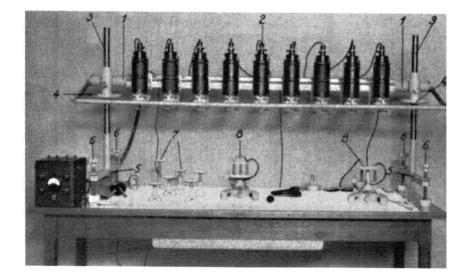

Figure 9.1   Zeiss multiplex equipment

Note: Number 1 is the bar from which the projectors (2) could be hung. Number 3 are the vertical supports for the bar. Number 4 enabled the adjustment of the height of the bar. Item 5 could be tilted by means of the screws (6) to tilt the projectors about the X or Y axes. Number 7 are the stages, adjustable in height to represent height and plan control points. Number 8 refers to the device used to make measurements and plot detail and contours. A horizontal white platen has a small hole in the middle, illuminated from below, which is the measuring mark. Vertically beneath the measuring mark is a pencil lead which is used for plotting. The platen can be moved vertically to view the model at different heights and has a scale attached for measuring heights.

Source: Author's own collection.

every couple of models along the strip. The whole strip could then be adjusted to fit all of the existing control and detail and the contours subsequently plotted. This technique was called 'bridging' because the photogrammetrist 'bridged' from one platform of control to another at the other end of the strip. This was the first effective method of aerial triangulation. In Britain, aerial triangulation was adopted by the Ordnance Survey from the late 1940s, but the time-consuming nature of the calculations meant that it was not widely adopted until computers came into general use from the late 1950s. With the application of computers, photogrammetry was better able to meet the demands of national mapping agencies and scientists who needed to make accurate non-contact measurements of surfaces.

Multiplex was rapidly adopted by the United States Geological Survey and used for mapping in conjunction with the Tennessee Valley Authority as part of the land reclamation work in the 1930s. It was also subsequently adopted in Canada for medium scale topographic mapping. Multiplex was adopted by both the United States and Canadian armies, and played a significant role in the mapping carried out prior to Operation Overlord in 1944, the allied invasion of Normandy. Multiplex was widely copied after the Second World War and became the standard equipment for much medium scale mapping into the early 1970s with organisations, such as the Directorate of Overseas Surveys (DOS), which undertook topographic mapping as part of Britain's overseas aid programmes. The only problem with multiplex was that it used reduced sized copies of the original aerial photography, meaning that it was really only suitable for medium scale mapping (roughly 1:50,000).

In the United States a similar system was developed by Kelsh that was capable of taking full-size aerial photographs (230 × 230 mm). Although widely used in the United States by both the US Geological Survey and private companies, it was little used outside of North America. The Kelsh plotter was used to produce the first practical orthophoto instrument, one that was capable of removing both the tilt and height displacement errors from aerial photography, to produce orthophotographs. Orthophotography and orthophoto maps have distinct advantages over conventional line mapping, since the resultant maps have all the information content of the original photography and the planimetric accuracy of the line map. Orthophotomapping was rapidly adopted by the US Geological Survey for mapping swamps, and was employed by the US Army for mapping forested areas in Viet Nam. Forestry agencies in North America also used it extensively. In Europe, the main early adopters were in Sweden for the Economic Map of Sweden, and North Rhine-Westphalia in Germany for 1:10,000 mapping. It was widely used in Australia for mapping the interior. In Britain, the Ordnance Survey use orthophotography in its revision of line mapping. The Directorate of Overseas Surveys used orthophotomapping for its maps of Aldabra, and in its series mapping of Botswana at 1:50,000 and 1:100,000, and of The Gambia at 1:50,000 in the late 1960s and early 1970s.

Using Photography: Photographic Interpretation and Geographical Analysis

While attention in a Western European context during the First World War largely focused on the use of aerial photography for topographic mapping and plotting enemy positions, the potential of aerial photography for mapping other kinds of information was quickly and more widely recognised. Skills honed in air photo interpretation of enemy positions were well suited to studying other objects in the landscape. Lieut.-Colonel George Adam Beazeley's presentation to the Royal Geographical Society in 1920 on the use of aerial photography in archaeology was based on his realisation that previously unidentified archaeological sites could be visible on photographs.[27] His ideas were important in interesting a new generation of archaeologists in the potential of aerial photography. Foremost among them in Britain was O.G.S. Crawford, who worked with the Ordnance Survey and, with Alexander Crawford Keiller, produced *Wessex from the Air* in 1928. Hugh Hamshaw Thomas was a further important advocate of the use of aerial photography. He had served in Egypt and Palestine with the Royal Flying Corps in World War I and with the Royal Air Force in liaison with the Royal Engineers. During the British-Arab advance on Gaza in 1917 he had aerial photography of the town undertaken, and, from that, created what is credited with being the first urban map from aerial photography.[28] In civilian life, Hamshaw Thomas was a biologist in Cambridge. In his use of aerial photography of the Near East, he realised that vegetation patterns in the Jordan Valley, which were impossible to discern on the ground, were clearly visible from the air.[29] There are parallels here with the work of Hugh Cott in World War II, the subject of Forsyth's essay, chapter 11. After 1918, Thomas became a keen advocate of the use of aerial photography in the environmental sciences.

Others also recognised that the techniques of aerial photography could assist in geographical exploration and research where landing or groundwork was difficult.[30] In Canada, for example, float planes and small flying boats to undertake aerial photography were widely used in Canada between the wars given the widespread availability of rivers and lakes.[31] Lack of suitable landing grounds restricted the otherwise successful use of aerial photography for geographical analysis in Ceylon,

---

27   George A. Beazeley, 'Surveys in Mesopotamia during the War', *Geographical Journal* 55 (1920): 109–23 [and see 'Discussion', pages 123–7].

28   Dov Gavish and Gideon Biger, 'Innovative Cartography in Palestine 1917–1918', *Cartographic Journal* 22 (1985): 38–44.

29   H. Hamshaw Thomas, 'Aircraft Photography in the Service of Science', *Nature* 105 (1920): 457–9.

30   P.R. Burchall, 'An Investigation of the Possibilities Attaching to Aerial Co-Operation with Survey, Map-Making and Exploring Expeditions', *The RUSI Journal* 67 (1922): 112–27.

31   S. Bernard Shaw, *Photographing Canada from Flying Canoes* (Burnstown, Ontario: General Store Publishing House, 2001).

but the Irrawaddy forest survey for the Forest Department in Burma, carried out by the Survey of India in 1923 and 1924 was more successful.[32] Although authors in the United States after 1918 advocated the use of air photos for studying urban areas, the first serious attempts to use air photos for geographical analysis did not take place until a decade later.[33] A number of projects initiated in the United States under the New Deal of the 1930s made extensive use of air photos. This included the analysis of agricultural land carried out by the Agricultural Adjustment Administration and the work of the Tennessee Valley Authority.[34] These Federal Government projects encouraged the development of the air photo industry and the adoption of air photo methods by other organizations.

The systematic use of aerial photography for exploration or geographical studies more widely was limited during the inter-war period. This was due in part to the limitations of technology, and, not unimportantly, to the cost, especially in the years following the Wall Street Crash of 1929. Again, the outbreak of the Second World War changed this situation: military demands drove technological advance which later had civil application. Germany was the first country to start the systematic collection of geographical data for military purposes, carrying out systematic reconnaissance flights from 1937.[35] The British were obtaining widespread coverage of potential targets or places of military significance from 1939.[36] Pre-war German efforts used large, heavy cameras in low flying aeroplanes to collect large-scale photography, whereas the British used faster, higher altitude flights, more like the kinds of sorties that would be later flown under combat conditions. German efforts largely focused on aerial photography for mapping and photogrammetric use, or for the support of troops, whereas the British and, later, the Americans, used aerial photographic work and its associated mapping more strategically, in bombing campaigns and in planning land campaigns.[37] In the British context, increased reliance was placed on the use of aerial photography

32   C.G. Lewis and H.R.C. Meade, *Air-Survey in the Irrawaddy Delta, 1923–24* (Dehra Dun: Survey of India, 1925).

33   Most notably Willis T. Lee, 'Airplanes and Geography', *Geographical Review* 10 (1920): 310–25, and W.L.G. Joerg, 'The Use of Airplane Photography in City Geography', *Annals of the Association of American Geographers* 13 (1923): 211; Vernor Clifford Finch and Robert Swanton Platt, *Geographic Surveys: Geographic Surveying* (Chicago: Geographic Society of Chicago, Publication No. 9, 1933); G. Donald Hudson, 'The Unit Area Method of Land Classification', *Annals of the Association of American Geographers* 26 (1936): 99– 112.

34   Mark Monmonier, 'Aerial Photography and the Agricultural Adjustment Administration: Acreage Controls, Conservation Benefits, and Overhead Surveillance in the 1930s', *Photogrammetric Engineering and Remote Sensing* 68 (2002): 1257–62.

35   Roy M. Stanley, *World War II Photo Intelligence* (London and New York: Scribner, 1981), 43.

36   Constance Babington Smith, *Evidence in Camera. The Story of Photographic Intelligence in World War II* (London: Chatto and Windus, 1958), 21–4.

37   Stanley, *World War II Photo Intelligence*, 51–6.

after 1940 and the retreat of Britain's armed forces from mainland Europe. Initially, aerial photography was used in target identification. Subsequently, aerial photography was used for the systematic study of German cities, identifying functional areas, industrial zones, and the productive capacity of factories as part of the planning for the bombing campaigns. In planning the Normandy landings of 1944, aerial photography was used to determine beach gradients, 'goings' (the suitability of terrain for the cross-country movements of different types of vehicles, something that was pioneered in North Africa), and to identify suitable sites for the rapid construction of airfields, beach defences and inundated areas.

After 1945, the western allies had large numbers of aeroplanes no longer needed operationally, together with cameras suitable for mapping and reconnaissance, and relatively large numbers of personnel trained in air photo interpretation and in the technologies of aerial photography while in the skies. The advent of the Cold War prompted further investment in the technical development of aerial photography and accelerated its exploitation. In Britain, the government took the view that the whole country should be provided with aerial photography, and that this was to be achieved by the late 1940s. The resultant photography was used to revise existing Ordnance Survey maps, and to produce photomaps for use by planners then undertaking the reconstruction of Britain's bombed cities and who could not – or would not – wait for the revised Ordnance Survey maps to be produced. The aerial photography of the late 1940s has subsequently proven invaluable to geographers and others in their assessment of landscape and measurement of environmental change.[38] A further consequence of the Second World War in the British context was that geographers trained in air photo interpretation went back to their academic jobs and introduced air photo interpretation into the curriculum of many university geography departments. The use of aerial photography also became part of the curricula of geology, biology and archaeology departments.

What may in a sense be thought of as the coming of age of air photo interpretation came in 1960 with the publication of the *Manual of Photographic Interpretation* by the American Society of Photogrammetry.[39] This work was important as it brought together for the first time a wide range of examples of aerial photography to study different aspects of the human and physical environments. A further post-war development was the establishment of the Directorate of Colonial Surveys (later the Directorate of Overseas Surveys) under Martin Hotine. From the start, the Directorate was to be almost entirely dependent on aerial photography for its

---

38   For example, W. Gordon Collins and P.W. Bush, 'Aerial Photography and Spoiled Lands in Yorkshire', *Journal of the Royal Town Planning Institute* 57 (1971): 103–10; Sarah Newsome, 'The Coastal Landscapes of Suffolk during the Second World War', *Landscapes* 4 (2003): 42–59; Robin M. Fuller, 'The Changing Extent and Conservation Interest of Lowland Grasslands in England and Wales: A Review of Grassland Surveys 1930–1984', *Biological Conservation* 40 (1987): 281–300.

39   Robert N. Colwell, *Manual of Photographic Interpretation* (Falls Church: American Society of Photogrammetry, 1960).

work. Initially, its work focused on the provision of basic topographic mapping for colonial or newly-independent African countries, or for use in the Americas and in South-East Asia. It was soon realised, however, that the development of those countries required other kinds of mapping, such as soils, geology, land use and vegetation, and other kinds of technical and institutional solutions. This realisation led to the establishment of the Land Resources Division (later the Land Resources Development Centre). The work of this Division was based almost entirely on the use of aerial photography. Workers in the Division were able to take advantage of the development by the Australian Commonwealth Scientific and Industrial Research Organisation (CSIRO) of land systems analysis. Following the threatened invasion of Northern Australia by the Japanese in World War II, the Australian government realised that it needed to develop the country's Northern Territory in order to keep it secure. The vast extent of the territory meant that using aerial photography was the only viable approach to surveying the territory's resources. Clifford Stuart Christian and George Alan Stewart realised that in a natural or semi-natural environment, with any given lithology and geological structure and a uniform weathering regime, only a narrow range of terrain types would develop. Each of the terrain units would also be characterised by particular assemblages of vegetation, range of soil types and hydrology. While most of these could not be easily interpreted on the photography, the terrain units and their vegetation would be. By interpreting the smallest units, called facets, it would be possible to define the extent of the larger units, which came to be called land systems.[40] Efficient field sampling could then be carried out to determine the soil types and vegetation types present. This approach, called Land Systems Analysis, was taken up by a wide range of organisations, including, in Britain, the Land Resources Division, the Transport and Road Research Laboratory, and the Royal Engineers. Its use survived the introduction of satellite remote sensing in the 1970s.

Conclusion

Almost immediately upon its development in the late nineteenth century, aerial photography seemed to offer the promise of a quick and efficient means for the collection of information on the human and physical environment. Realising that promise required, however, that several conceptual and technical problems had to be overcome. This chapter has shown that, whilst there was an element of common origin in the effects of two world wars upon advances in air-borne photographic technology, the technological approaches employed in solving those problems differed by country. Most Anglophone countries initially focused on simple, if labour intensive, approaches. In Britain, technological advances were

---

40   C.S. Christian, S.T. Blake, L.C. Nokes, and G.A. Stewart, 'No. 1 General Report on Survey of Katherine-Darwin Region, 1946', *Land Research Surveys* (Melbourne: Commonwealth Scientific and Industrial Research Organization, 1953).

mainly driven by military requirements for rapid map-making in the field. In mainland Europe, the preference was for technical solutions which, on the whole, employed more complicated and expensive equipment and the services of highly-skilled operators. Both approaches continued in parallel until the 1960s, when instrumental photogrammetry became the shared norm. With the normalisation of instrumental photogrammetry from this period, the metrological use of aerial photography became the exclusive domain of technical experts, largely in national mapping agencies. In that regard, expertise that been developed in consequence of shared military imperatives was more readily and widely developed in civilian institutions and, to a lesser extent, in universities.

The use of imagery itself was not technically demanding. It often required nothing more complicated than a simple stereoscope and, for this reason, the application of aerial photography continued to be employed, and developed further, by a wide range of disciplinary practitioners such as environmental scientists, landscape historians, geographers, and archaeologists, to name only a few. Perhaps because interpretative questions and related applications were relatively technologically simple, aerial photo interpretation became and has remained a more evidently democratic part of the development and use of aerial imagery. In the early twenty-first century, advances in digital photogrammetric techniques have, if anything, significantly decreased the level of skill necessary to comprehend and interpret the images produced from aerial photography or, in a related context, satellite imagery. This fact has opened up, of course, the prospect for wider participation in map making and for a 'participatory GIS' well beyond that already offered by crowd sourcing.

Behind these questions of access, however, and behind the facility with which commonly available aerial photography is read and interpreted, are a set of technological challenges and their associated innovations and solutions. These were initiated by the demands of the military: in general terms, aerial photographic technology is a product of knowing where and how to wage war. These technological innovations began with stabilising the flight of the aircraft itself. Only later did they involve fixing and regulating the camera and the films used in what, commonly, was a fast moving and vibrating device which could not guarantee that the image produced of the earth beneath was directly vertical. The fact that truth from the air did not always correspond with truth on the ground was the very *raison d'etre* of aerial photo interpretation. For the air-borne devices themselves – aircraft, pilots, cameras variously attached and refined – the technological development of aerial photography was influenced by the terrain circumstances of different parts of the world. For these reasons, it is helpful to understand the technological questions and solutions behind aerial photography not as universal shared solutions easily wrought but as the product of particular institutional, physical, and interpretive human geographies.

Chapter 10

Uncovering Camouflage: The Technology of Location and the Craft of Erasure

Isla Forsyth

Military camouflage is a technology and an aesthetic which attempts to disrupt and to disturb. It first began to be systematically used by the British armed forces in World War II and was so in response to technological advances in the aeroplane as a weapon and to ever greater accuracy in targeting. Although the primary purpose of military camouflage is that of deception, there has been little sense of any unease surrounding the ambiguous nature, and at times, sinister character of camouflage. The predominant explanatory narrative is one of the ingenuity of artists in association with the efficiency of soldiers; camouflage is in this sense seen as a fantastical sleight of hand, the tree that holds a soldier on lookout or a city protected by a mask of seeming bomb damage. This unproblematic reading of camouflage may offer an interesting window on British military ingenuity, but taken at face value it offers too simple, even too certain, an account. In her studies of military geographies, Rachel Woodward has warned against questions of interpretative complacency or negligence when examining the geographical presence of the military and the militarism of culture; '[m]ilitary geographies may be everywhere, but they are often subtle, hidden, concealed or unidentified'.[1] Here I explore aspects of the geography of militarism by looking at objects and practices which were intended to be subtle, hidden, and, by virtue of the technologies of their representation, to remain concealed or unidentified.

This chapter explores the establishment and training by the British military authorities of its first batch of Camouflage Officers (or 'camoufleurs') in World War II. What follows examines how the entwining of visual literacy, geographical knowledge, artistic flair, and scientific rigour produced a subversive technology and a subtle instrument of war. The chapter pays tribute to the ingenuity of camouflage's practitioners, but it also works to distort the prevailing narrative of camouflage by making visible its history as a mercurial aesthetic – one which was designed in response to technologies of surveillance and targeting – and one which, until now, has evaded sustained scrutiny. The chapter begins with a particular but important instance in the development of camouflage as a technology of visual

---

1 Rachel Woodward, 'From Military Geography to Militarism's Geographies: Disciplinary Engagements with the Geographies of Militarism and Military Activities', *Progress in Human Geography* 29 (2005): 718–40, quote from page 719.

deceit. From this example I address some of the broader connections between observation, the technologies of viewing, and the practices of camouflaging. I do so with particular reference to the work in Farnham, England, of the Camouflage Development and Training Centre from October 1940 and to the ways in which the individuals there gathered techniques and technologies designed, in particular, to deceive the airborne bomber. This is an essay about the visual art and craft of technological and military deception, an art and a craft which, even as it aimed to mislead the eye, depended vitally on understanding the geographical settings of the objects concerned.

Hugh Cott's Camouflaged Gun

In the early summer of 1940 at an undisclosed location on the east coast of England, Dr Hugh Cott, a Cambridge zoologist was working on a railway siding (Figure 10.1). The British military authorities had charged Cott with setting up a comparison, in terms of their camouflage, between two railway guns; one demonstrated standard British military camouflage design; the other with camouflage which mimiced adaptive colouration and patterning in nature. Cott was pleased to find distraction from academic life in order to work with the military. In a letter to a colleague he wrote that 'Zoology seems rather to sink into insignificance these days, beside the events happening elsewhere. Cambridge is quiet and normal and here I can almost forget the war – but that is not so much to my liking'.[2] Cott was eager, even desperate, to be so utilised by the British armed forces in the development of camouflage technology. Cott had served in the British Army in Ireland in the 1920s and, while there, had developed a deep interest in understanding the means by which animals disappear into and merge with their natural surroundings. He pursued this interest at university where he studied the question of adaptive colouration in nature for nearly two decades, conducting numerous expeditions in South America, Africa and Europe. The results of this work rewarded him with some of the secrets of biological camouflage. From careful observation in the field, Cott explored the biological camouflage techniques of concealment, visual disruption and display, and established himself as a leading figure on the subject. Observation of nature and of the capacity of natural creatures and plants to 'hide' themselves, even in movement, from the observing eye, would lead him to make connections to human cultures and to the world of war. From his fieldwork, Cott noticed that there appeared to be 'a close analogy between the adaptations of animals and the inventions of man – whether used in nature or in battle. Indeed, almost every invention, including some of the most recent,

---

2   Glasgow University Archives [hereafter GUA] DC6/741, Letter from Cott to Kerr, 3 April 1949.

Figure 10.1 Cott's camouflage gun experiment
Source: University of Glasgow Archive Services, John Graham Kerr collection, GB0248 DC6/780.

has its counterpart in the modifications and behaviour of various wild creatures'.[3] With his findings published in scientific papers and as the result of his seminal text *Adaptive Coloration in Animals* (1940), Hugh Cott became established as an authoritative figure on biological camouflage.

Throughout *Adaptive Coloration in Animals*, Cott drew analogies between the facts of concealment in nature and the techniques of visual obfuscation in culture (the net curtain, the pickpocket). He drew, in particular, close parallels between

---

3 Hugh Cott, 'Camouflage', *The Advancement of Science* IV (1948): 300–12, quote from page 301.

surviving and thriving in nature and in modern warfare.[4] In his view, 'the primeval struggle of the jungle, and the refinements of civilized warfare, have here very much the same story to tell'.[5] Cott further believed that the issue of camouflage had in the main been ignored by the military and that this might have potentially serious consequences for any future conflicts: 'Camouflage, a product of the Great War, is still in its infancy – a child suffering from arrested development. Its importance and possibilities have yet to be fully appreciated'.[6] The invitation in the early summer of 1940 to demonstrate the potential effectiveness of biological camouflage to the British military was thus most welcome. Cott hoped that his camouflaged railway gun would provide the military with proof that camouflage design and technology adhering to scientific principles could be an effective means of concealment.[7]

From his experiences in the field, Cott understood that in undertaking the task to camouflage a gun effectively, he would need to mimic the techniques employed in nature. As he commented: 'The conditions of light which affect the appearance of a snake are the same as those which cause a gun or torpedo-tube to stand out conspicuously *even against a background covered with exactly the same paint*. Hence it is essential that rounded objects be treated with paint properly graded in tone so as to counteract the effects of relief'.[8] Cott had made a study of nature's colours as they were enrolled in interspecies warfare. For the art historian Diana Young, colour is both knowledge and being, something that is experienced and shaped by cultures and histories of interpretation and representation.[9] For Hugh Cott, the colourful patterning on non-human animal bodies, their contrast, and their blending in relation one to another and in relation to the ecological dialogue with their habitat could aid in matters of human survival if applied in a military context. Cott's gun and his painting of it is, therefore, an intriguing object of significance well beyond the military technology itself. His work upon it brought the natural and military world into correspondence. Study of the Cott gun offers insight into the production of technological innovation as a performance of human and nonhuman (animal and technology), a practice that Andrew Pickering calls 'a performative image of science', the process whereby scientists 'manoeuver in a field of material agency'.[10] Through Cott's scientific efforts and artistic eye, the

---

4   Isla Forsyth, 'The Practice and Poetics of Fieldwork: Hugh Cott and the Study of Camouflage', *Journal of Historical Geography* 43 (2014): 128–37.

5   Hugh Cott, *Adaptive Coloration in Animals* (London: Methuen, 1940), xi.

6   Ibid., 438.

7   GUA, DC6/753, Letter from Cott to Kerr, 19 August 1940.

8   Cott, *Adaptive Coloration*, 46. Original emphasis.

9   Diana Young, 'The Colour of Things', in *Handbook of Material Culture* (ed.) Chris Tilley, Webb Keane, Susanne Kulcher, Mike Rowlands and Patricia Spyer (London: Sage, 2006): 173–85.

10   Andrew Pickering, 'Cyborg History and the World War II Regime', *Perspectives in Science* 3 (1995): 1–49, quote from page 7.

natural pigment and patterns of the snake were replicated by human-made paints and materials on human-made weapons. Used in this context, colour, form, and pattern traversed the boundaries of science, militarism, nature, and art. Both of the guns, the one conventionally attired in then standard military patterns and Cott's more 'natural' weapon, made use of the same paints, earthy muddy pigments of greens and browns, and contrasting tones of black and white. The difference between them lay in the different techniques of application and in their differing patterns. Cott was at pains to show that the same colours on the two guns could be used to produce quite different effects with regard to the visual technologies of warfare.

Upon completing his experiment, Cott enthused: 'My gun is quite indistinguishable while its counterpart can be easily recognised'.[11] In numerous flights to test the effectiveness of the different schemes of camouflage, the gun demonstrating official military camouflage was identified each time and described as being as 'glaring as a stick of peppermint rock'.[12] Cott's gun, on the other hand, seemed simply to melt into the backdrop; it was identified only once, on a low flight path when its exact coordinates were anyway known to the pilot. For Cott, his gun experiment was a victory in the campaign to promote the scientific principles of camouflage. Cott contemplated the potential and broader role camouflage would have in the war if effectively applied: 'By its successful employment it is possible not only to save lives, material and equipment for subsequent use, but also to bewilder the enemy, causing him to make false moves, to waste his bombs and ammunition, and to dispose his forces in the way best suited to our plans'.[13] Cott recognised through his gun experiment the potential of camouflage to deceive the eye and, in turn, to deceive the body of the enemy into making mistakes.

Aerial Visualisation and Military Observation

Jonathan Crary has discussed how vision is 'inseparable from the possibilities of an observing subject who is both the historical product *and* the site of certain practices, techniques, institutions, and procedures of subjectification'.[14] The practices of observation are embedded within and shaped by spatial and temporal contexts. In turn, the techniques of observation have an influence upon the spatial and temporal setting of the observer. This can be illustrated by examining the nature and the consequences of military observation. The ways in which vision is deployed in and by the military are influenced by a range of instruments, technologies, knowledges and social and historical contexts. Fraser MacDonald

---

11   GUA DC6/753, Letter from Cott to Kerr, 19 August 1940.

12   GUA DC6/754, Letter from E.D. Swinton to Sir Graham Kerr, 27 September 1941.

13   Cott, 'Camouflage', 300.

14   Jonathan Crary, *Techniques of the Observer: On Vision and Modernity in the Nineteenth Century* (London: MIT Press, 1990), 5.

has explained that 'looking' in military terms is performative: 'to have a target in sight is to have already changed the relation between subject and object. The technology of optics, from the earliest field telescope to modern systems of radar and optoelectronic surveillance, can arguably be reduced to the triumph of speed and the defeat of proximity'.[15] Thus, as the technologies and instruments of vision have developed, military authorities have effectively already implemented strategies and technologies with regard to targeting. Camouflage thus reveals – ideally, by not revealing itself – what we may think of as a competitive tension within the military since, as instruments of visual acuity develop in their accuracy and potency and as the speed and mobility of violence in war increases, military operatives need also to have the techniques to mitigate these developments.

Since the First World War the aeroplane in Western Europe has been deployed both for surveillance and for bombing purposes (see also Collier's chapter in this volume). In turn, camouflage has been adopted and adapted in order to create visualities of deception to undermine these aerial technologies. In the First World War, the artist Solomon J. Solomon helped establish the British military's initial camouflage efforts, advising on technical issues, recruiting artists who could help to devise camouflage schemes and overseeing a site near the Western Front for camouflage experimentation.[16] Solomon studied aerial reconnaissance photographs in detail in order to become acquainted with aerial visualities and he developed a number of methods to undermine enemy observers. To aid his work, Solomon bought a radiographer, a machine which projected images on to a screen and which, when used with mirrors and lights, could be employed to study aerial photographs in considerable detail.[17] Solomon's early work with regard to the aerial view and attempts to disrupt it prefigured the concern of the British armed force with camouflage in World War II.

It is important to note, however, that the idea of the 'view from the air' has wider civic significance that just an association with a technology linked to warfare. The aeroplane was quickly incorporated in a complex moral geography of citizenship and nationhood; 'airmindedness' was fostered in different places through different means. As geographer Peter Adey explains, air-mindedness was used by nation-states to produce 'a body readied for performance'.[18] The idea of airspace understood from a military perspective is, as Alison Williams

---

15   Fraser MacDonald, 'Geopolitics and 'The Vision Thing': Regarding Britain and America's First Nuclear Missile', *Transactions of the Institute of British Geographers* 31 (2006): 53–71, quote from page 54.

16   Olga Phillips, *Solomon J. Solomon: A Memoir of Peace and War* (London: H. Joseph, 1933).

17   Nicholas Rankin, *Churchill's Wizards: The British Genius for Deception 1914–1945* (London: Faber and Faber, 2008), 256.

18   Peter Adey, '"Ten Thousand Lads with Shining Eyes are Dreaming and Their Dreams are Wings": Affect, Airmindedness and the Birth of the Aerial Subject', *Cultural Geographies* 18 (2010): 63–89, quote from page 83.

has discussed, not one of simple paths, trails or corridors of air through which aircraft pass, but, rather, one of distinct spaces of performance, each operating as the projection of power.[19] In essence, since World War I and through the agency of aircraft, military power has 'stretched' into the vertical.[20] Williams contends that, by treating military aerial spaces as control over space and control above space and by identifying the multiplicity there of military performances through surveillance, defence, and offensive operations, space can be understood as 'a volume rather than a flat bound plane'.[21] The projection of power therefore extends horizontally across and through space following the path of flight, but also moving vertically along and through trajectories of ordnance and surveillance. In this interpretation, military airspaces become more active and open to research, more dense, thick with the politics of power, control, and connected, as I hope to show, to the terrestrial geographies of military camouflage and the technologies of visual subversion.

The history of military camouflage development can be understood as the transference of biological science to military practice in addition to being seen as the product of a relationship between carefully staged visibility, methodical observation and the technology of deceit. In this context, Helen Macdonald has explored the entangled and embodied relationship of birdwatching and aircraft spotting in mid-twentieth-century Britain. She suggests that during this period particular versions of ecological, national, and social identity were conveyed through the practice of observation.[22] The observer's body was keenly trained, their eyes sharpened to recognise and identify, and it was a practice that produced change. The body of the bird was reconstructed through the interpretation made by the human eye and specific observation practices, which, in turn, acted to produce the embodied observer. In drawing comparisons between the new birdwatcher and aircraft observers, Macdonald argues that, in the 1940s, the activities of the birdwatcher and the aircraft spotter or observer became connected; the observer, whether of bird or of plane, possessed a knowledgeable gaze. Theirs was a visualisation of ecology that had authority which both influenced and was influenced by militarism (in the use of binoculars, clothing designed to blend with the environment, charts of type specimens and so on). Military camouflage could

---

19   Alison Williams, 'Reconceptualising Spaces of the Air: Performing the Multiple Spatialities of UK Military Airspaces', *Transactions of the Institute of British Geographers* 36 (2011): 253–67.

20   Alison Williams, 'Flying the Flag: Pan American Airways and the Projection of US Power Across the Interwar Pacific', in Fraser MacDonald, Rachel Hughes and Klaus Dodds (eds) *Observant States: Geopolitics and Visual Culture* (London: I.B. Tauris, 2010): 81–99, quote from page 84.

21   Williams, 'Flying the Flag', 256.

22   Helen Macdonald, '"What Makes You A Scientist Is The Way You Look at Things": Ornithology and the Observer 1930–1955', *Studies in History and Philosophy of Biological and Biomedical Sciences* 33 (2006): 53–77.

be seen similarly to enmesh the ecological observer and the military specialist as it produced new visualities by drawing from art and science in order to transform military technology. Camouflage is thus something of a 'visual curiosity' born through entwining visual culture and practices of observation in order to bewilder, deceive and, ultimately, go unnoticed.[23]

Anticipating Shock and Awe

Because of technological developments in weaponry, camouflage was directly associated with the aeroplane and with the camera before 1939. Yet it is clear that World War II represents an unprecedented phase in camouflage's geography and history, as its techniques were continually adapted in order to find novel ways to outwit enemy eyes. Even as this is true, however, it is also the case that accounts of camouflage's development in World War II have been to view the subject as rather benign and without active agency. One interpretation is that military camouflage was innovated and implemented by a cast of eccentric artistic individual and a few scientists.[24] Another interpretation has it as a story of the ingenuity of the military in pooling a diverse group of military specialists and outside experts to develop deception techniques.[25] While these narratives highlight something of the technological development of camouflage they have somehow, in focusing on its origins, also allowed camouflage itself to avoid its own scrutiny, to be perceived as a technology with little or no ethics or politics; a technology divorced either from its social construction or read without reference to its association with violence. Camouflage was, as this chapter shows, intimately linked with air power and with the threat of bombing.

In the 1930s, as tensions between the European nations grew more and more fraught, the British military began to prepare for possible future conflict. A large part of the military's role is involved not with waging war but in preventing it, especially attacks upon vital infrastructure. Considerable effort was expended therefore in the preparations for defence.[26] One of the greatest fears in Britain was that of airborne invasion. By the late 1930s, the aeroplane was cast as a 'technological triumph', something which had essentially redefined the nature and methods of war.[27] For Ken Hewitt, as Western militaries realised the importance

---

23   MacDonald, 'Geopolitics and "The Vision Thing"', 68.

24   See Roy Behrens, *False Colors: Art, Design and Modern Camouflage* (Iowa: Bobolink Books, 2002); Henrietta Goodden, *Camouflage and Art; Design for Deception in World War 2* (London: Unicorn Press, 2007).

25   Rankin, *Churchill's Wizards*, 325–33.

26   Michael Bateman, and Raymond Riley, *The Geography of Defence* (London: Croom Helm, 1987).

27   Daniel Pick, *War Machine: The Rationalisation of Slaughter in the Modern Age* (London: Yale University Press, 1993), 73.

of the aeroplane in conflict, they also become aware of urban centres being vital yet vulnerable sites for the state.[28] At the beginning of the war, the British military began to calculate the potential impact of the aeroplane as bomber. As Patrick Deer puts it, the military authorities recognised the 'capacity for the bomber to outrun naval or land forces in attacks on enemy airfields, factories, or cities'.[29] The British realised that aerial attacks on cities were likely and that the prospect of aerial bombing of cities promised destruction and violence on hitherto unprecedented scales. The aeroplane's blurring of vertical and horizontal battlespace paralleled the blurring of the division between civilian and target. As the city, through a series of visualizations, became a target, civilian populations in World War II became justifiable casualties; the aeroplane created a 'new and terrible space of war on the Home Front'.[30] As Derek Gregory has highlighted, World War II was an important 'moment' in defining the geography as well as the history of bombing.[31] In the face of aerial warfare, the British military, itself embracing bombing as a means of conflict from above, sought new methods such as camouflage to mitigate and protect against these terrors. As surrealist artist and camoufleur Roland Penrose explained in *Home Guard Manual of Camouflage* (1941), 'the aeroplane among other things is the eye of the modern army. Its invention makes camouflage more urgent, more difficult'.[32]

Yet it is also the case that Britain's military authorities were unaccustomed to employing concealment as a technology of conflict. Some personnel viewed it as underhand, or, even, a cowardly means of conflict. Despite such misgivings, the British authorities took pains to draw together experts from the field of aesthetics and visuality and specialists from the arts, sciences, film, and stage with the aim of forming new networks of military personnel. The stated intention was to make these non-military experts understand modern ways of fighting in order that they might develop a visual technology for warfare, and so, in their working and thinking, become 'camoufleurs'.

Training to Disappear and the Technology of Concealment

In October 1940 a 'curious collection' of men disembarked from a London train at Farnham, Surrey. The group, which included Hugh Cott, was the first and

---

28   Ken Hewitt, 'Place of Annihilation: Area Bombing and the Fate of Urban Places', *Annals of the Association of American Geographers* 73(1983): 257–84.

29   Patrick Deer, *Culture in Camouflage: War, Empire, and Modern British Literature* (Oxford: Oxford University Press, 2009), 73.

30   Ibid., 74.

31   Derek Gregory, 'The Rush to the Intimate: Counterinsurgency and the Cultural Turn in Late Modern Warfare', *Radical Philosophy* 150 (2008): 8–23.

32   Roland Penrose, *Home Guard Manual of Camouflage* (London: G. Routledge & Sons, 1941), 7.

perhaps the most illustrious students of British military camouflage to attend the Camouflage Development and Training Centre (CD&TC), Royal Engineers (RE). Members of the group had been handpicked for their work in camouflage because of their varied and diverse professional expertise. The arts were represented by surrealist painters Julian Trevelyan and Roland Penrose, designers Steven Sykes and Ashley Havinden, as well as by stained-glass artists, filmmakers such as Geoffrey Barkas, a cartoonist, a surrealist poet, and a restorer of religious art.[33] The cast included the well-known stage magician and conjuror Jasper Maskelyne. Other disciplines were also drawn upon, including men with expertise in architecture and in engineering. Cott was the only scientist and the only person on the course with military experience.[34] At Farnham, the professional skills of these individuals would be subsumed and merged as they were made specialists in camouflage. Their brief as 'camoufleurs' was to 'propagate the camouflage gospel and train unit camouflage instructors in formations in Britain and overseas, whilst also advising on all necessary camouflage designs and technology'.[35]

A normal training day at the school lasted from 9 am until 5.30 pm, and consisted of a mix of tutoring, training exercises, and experimentation. In the mornings, the students, as well as participating in basic drill training, attended lectures. The topics of these lectures ranged across the basic principles of camouflage: effective camouflage in nature; the importance of background; the visibility of an object; the interpretation of air photographs and 'what they reveal'.[36] The nascent camoufleurs were taught the practicalities of carrying out camouflage schemes including camouflage materials; the general application of camouflage in the field; and the specialist application of camouflage to medium artillery positions. The lectures were complemented by practical exercises in the afternoons which were conducted in the castle's grounds. These included the trainees attempting to conceal both themselves and military equipment against the natural surroundings, and testing the design of dummy aeroplanes which had been constructed in the camouflage workshops, formerly Farnham Castle's stables. For some men, such as the artist Julian Trevelyan, this period and programme of activities proved to be highly agreeable. In his memoir *Indigo Days* (1957), he recalls it as an enchanting rural idyll with days of pearly autumn mornings in the wooded precincts of Farnham Castle.[37] For others, such as the magician Maskelyne, the time spent in learning techniques of concealment was not, initially at least, very fulfilling. 'Six

---

33   Geoffrey Barkas, *The Camouflage Story (From Aintree to Alamein)* (London: Cassell, 1952).

34   GUA DC6/758, Letter from Cott to Kerr, 26 October 1949.

35   D.J.C. Wiseman, *The Second World War 1939–1945 Army Special Weapons and Types of Warfare Vol. III Visual and Sonic Warfare* (London: The War Office, 1953), 152.

36   Dean Gallery Archive, Edinburgh [hereafter DG] GMA A64. Details for the CD&TC are compiled from material in the James McIntosh Patrick Archive in the Dean Gallery Archive.

37   Julian Trevelyan, *Indigo Days* (London: McGibbon and Lee, 1957), 116.

weeks of being told very elementary truths about the art of hiding things almost drove me out of my mind. I think I may say without particular vanity, and certainly without any aspersion on my lecturers, that a lifetime of hiding things on stage had taught me more about the subjects than rabbits and tigers will ever know'.[38]

As the first CD&TC course progressed, Farnham became a heterotopic, interdisciplinary, and dialogic site, one that created its own possibilities for the potentials of military camouflage to be realised. In these emergent dialogues, Cott appeared, initially, to cut a rather ambiguous and isolated figure amongst the colourful artists and artisans. His expertise and its particular application to warfare was soon recognised by his fellow camouflage trainees, however, and Julian Trevelyan's record of his work with Cott is insightful of the environment of collaboration fostered at the CD&TC: 'There was also a distinguished Cambridge zoologist, Dr Hugh Cott, who had written the most authoritative study on the protective coloration of animals and who now applied the principles he found in the animal kingdom to the disguise of guns and tanks. We laughed at him for his passionate addiction of counter-shading, the trick by which, for instance, the white belly and dark back of a gazelle, when seen at a distance in strong light, seem to flatten out and destroy form'.[39] This technique – of deception by design – would later become central to the lessons that Trevelyan taught as a Camouflage Officer:

> Occasionally I would be asked to give a demonstration of how to paint some piece of equipment so as to merge it with the broken country around it. I would arrive at the barrack square with pots of paint and brushes, and set to work daubing the shield of some anti-tank gun with spots of different greens and browns, touching the underside of the barrel itself with pure white on the principles of Dr Cott's gazelles. Against the dreary barrack walls it looked an unholy mess, but when it was wheeled out into the country and placed against the hedge, there were cries of astonishment at my magic.[40]

The technique used colour, texture and positioning to create a visual effect that disrupted sight and deceived vision. This example of a seemingly simple paint job made by its situation to appear something else provides an illuminating example of the accomplishments of the first batch of pioneer camouflage officers. Farnham and the CDTC proved an open space, geographically and socially, for the blurring and transformation of knowledges and techniques. Replicating the concealment technique that Cott developed in his railway gun experiment, Trevelyan's method resulted from understanding how 'Cott's gazelles' (namely biological camouflage) was coupled with his own artistic flair. This engagement with different disciplines and the co-mingling of knowledges and practices to create innovative camouflage designs and technologies even came to be appreciated by the most self-assured

---

38   Jasper Maskelyne, *Magic Top Secret* (London: S.I. Paul, 1949), 17.
39   Trevelyan, *Indigo Days*, 118.
40   Ibid., 130.

of camoufleurs. Even Maskelyne was convinced that his 'highly expert and specialised knowledge' in the art of hiding things could, after all, be improved upon by the assimilation of other expert knowledges. One instructive lesson, which he drew upon throughout the war, is revealed in his World War II memoir *Magic Top Secret* (1949). Through a description of inventing a concealing disguise for ships in the Middle East, he explained how 'We must use the magic known to a few specialists in the world of eye-doctors, of creating lines and projections, and patches of light and shade ... I had many interviews and conversations with such men, and with naval engineers, as well as artists and such people as Professor Cott, the Cambridge biologist and author of a famous work on animals' protective colourings, before I was satisfied enough to go ahead, first on scale models, and finally on a real launch'.[41] Maskelyne had to concede that creatures such a rabbits and tigers did have a thing or two to teach him about the craft of camouflage: 'In much of our work we had to go back for lessons and ideas to Nature, which, in the end, teaches man everything. We had to copy the colourings and also the mannerisms of animals and fishes and birds, which render themselves invisible'.[42] At Farnham, it soon became understood that through patient study of natural history and by attending closely to the natural environment of the battlefield, effective camouflage technology could be designed.

At Farnham, weapons of war such as anti-tank guns were aesthetically altered by the mutual constitution of art, science, design and military know-how. Each was brought into close correspondence, 'locked together as a complex, social, material, and conceptual cyborg entity'.[43] Through the ambiguous nature of camouflage, the CD&TC produced at Farnham a heterogeneous performance space, an ambiguous site of hybrid knowledge production and technological innovation. In setting up a dialogic space at Farnham Castle, the CD&TC and the fieldcraft practices of camouflage can be seen to form a regime which both allowed for, and made possible, 'disciplinary and material transgression'.[44] One important way in which this was done was by pooling skills and knowledge in order to produce innovative camouflage technology that had the potential to outwit airborne technologies of surveillance and targeting. For Ben Anderson, 'if the event is pre-empted, it is not simply prevented. Instead through sovereign action, the consequences of an event are brought into the present to be acted over'.[45] Bringing probable future threats (such as bombing raids) into the present in the artificial safety of the training ground in order that they might be understood and undermined was a key aim of

---

41   Maskelyne, *Magic*, 113.

42   Ibid., 157.

43   Pickering, 'Cyborg History', 18.

44   Trevor Barnes and Matthew Farish, 'Between Regions: Science, Militarism, and American Geography from World War to Cold War', *Annals of the Association of American Geographers* 94 (2006): 807–26, quote from page 809.

45   Ben Anderson, 'Hope for Nanotechnology: Anticipatory Knowledge and the Governance of Affect', *Area* 39 (2007): 156–65, quote from page 159.

the CD&TC. For Anderson, technologies undergoing experimentation are neither real nor imaginary but are, instead, in a process of becoming whereby the material fabric, so to say, is imbued with moral hopes, and, of course, by expressions of fear and anxiety. In World War II, the camouflage innovations developed by Cott, Trevelyan, and Maskelyne and other trainee camoufleurs can be understood in these terms. They worked to invent a technology of hope (for immediate personal safety and collective final victory) by undermining, even eradicating, actual threats; hope was built on the foundations of visual deception.

The camoufleurs quickly came to realise that the visuality inherent in bombing – in which camouflage was implicated – was vitally linked to aerial reconnaissance. Their task thus increasingly demanded technologies of camouflage to bamboozle both the observer and the bomber. Camouflage became enrolled in a contest between the bomber as air-borne destroyer and the grounded camouflage deceiver. As the camoufleur grew ever more expert in concealing military and civilian targets, the aerial observer was required to become ever more precise in deciphering the earth's surface, and the aerial bomber ever more proficient in targeting and attacking. It was vital, therefore, that the camoufleurs understood aerial perspective if they were to master the craft of concealment. Roland Penrose explained the camoufleurs' task in this regard: 'it is useless in warfare to be merely brave, if bravery means presenting oneself as a useless target to the enemy. It is far better to employ intelligence and concealment, so as to induce *him* to present a target. A man who is concealed can bide his time, watch for the enemy to expose himself and hold his fire until his target is sufficiently close to make sure of it'.[46] In order, then, that the British military camoufleurs should understand this aerial view they took to the skies.

Undermining Aerial Technologies of Observation

The emergence of aerial reconnaissance in conflict can be traced to its use by balloon observers on a limited and localised scale in the American Civil War and in the Franco-Prussian War.[47] These aerial observers were stationary, tethered to one location; accordingly their scope on enemy activity was partial and restricted. Powered flights transformed aerial reconnaissance because the aeroplane, in association with the air-borne camera, opened up and freed spaces for observation. These changes in aerial observation were begun to be seen in Western Europe in World War I. As Kitty Hauser explains, trench warfare, the dug-in and predominantly static nature of battle in the Great War, meant that the potential for victory 'depended more and more upon accurate mapping of the precise position

---

46 Penrose, *Home Guard Manual*, 102.
47 Roy Stanley, *To Fool a Glass Eye: Camouflage versus Photoreconnaissance in World War II* (Shrewsbury: Airlife Publishing, 1998).

of the enemy and its developing defences'.[48] Aerial photography thus instigated a new overhead technique to render the battle space and trench terrain visually legible; military aerial intelligence became an effective and potentially deadly weapon. It was the aeroplane (as well as the tank) that 'rescued the First World War from stalemate'.[49]

The requirement to conceal military intentions and their physical presence and strength of forces from the detached air-borne roving eye required that camoufleurs possess not only the technical ability to decipher aerial images, but also the expertise to compose a terrestrial image that could lie to the skilled eye, even one which, in safety and over time, might study the images in detail. The military needed camoufleurs to become experts in understanding the eye of their observant enemies in order that they could devise a scheme to mislead and deceive. As a military review of camouflage in World War II stated: 'training in concealment was inseparable from training in observation: the practice of one invariably demanded a knowledge of the other'.[50] Cott and his fellow camoufleurs realised that from above things on the ground appeared in seeming patterns. As one landscape painter and camouflage instructor told new recruits in his lectures, 'Everything falls into a patchwork quilt effect'.[51] Yet the camoufleurs had little time to appreciate the aesthetics of the earth from on high: 'to the Camoufleur the appreciation of these designs is more a practical issue; for if he is to effectively hide anything from a slit trench to an aircraft factory, he must learn not to disturb, or at least, to recreate, the basic pattern of the country'.[52]

The first thing to be concealed was the 'military signature' which revealed all to an observant enemy. As camouflage training literature explained: 'The army treads down the rich, natural texture and writes in white reflected light'.[53] Military lines of communication, camps and equipment betrayed their presence to the aerial cameras. These were the 'signatures' that the camoufleurs had to conceal. The training film, *Air View* (1941), emphasised this point by way of a dramatization involving the tracks of two careless soldiers who unthinkingly took a short cut across grass rather than fitting with the contours of the field boundary. Their lines of movement were clearly revealed in the aerial view. The white line of their paths, caught in reflection by the camera, disclosed their battalion's position to the enemy. The camp, an easy target, was attacked. Such careless actions, the

---

48 Kitty Hauser, *Bloody Old Britain: O.G.S. Crawford and the Archaeology of Modern Life* (London: Granta 2008), 29.

49 Ibid., 37.

50 Wiseman, *Visual and Sonic Warfare*, 167.

51 DG GMA, A64/1/16/3/1/8, The Training Problem Presentation [No date].

52 Julian Trevelyan, 'Camouflage', *Architectural Review* 96 (1944): 68–70, quote from page 70.

53 DG GMA, A63/3/1/4, Camouflage Instruction Chart No. IIIc Air Photographs Texture.

film exhorted, 'could cost the lives of many. It's up to you'.[54] In order to obliterate signs of the military's ground signature, the camoufleurs had to work with and blend into the pattern or rhythm of the background on which the military were operating. This, however, was not easy. Training literature emphasised the pressing need to consider the earth's patterns from the air view in order to camouflage effectively. As James McIntosh Patrick, landscape painter turned World War II camoufleur instructor at the CD&TC, emphasised in his lectures: 'We must become background conscious. Concealment in short, means fitting or merging something or oneself into the background and becoming part of the background. The machine-made soldier does not do this naturally'.[55]

This need to have objects blend and merge inconspicuously with their background meant that the camoufleurs had to understand the earth's surface not only from an aerial perspective but also through its colours, tones, and angles all of which the aerial camera would capture. The earth as captured by the camera working in black and white film was depicted in a 'series of tones ranging from black to white'.[56] Landscapes exposed through light and shade, shown as tones of grey, meant it was texture, shine, and shadows which told the earth's story from the air. The air-borne camera, it was realised, was dangerously obtrusive. It offered an unfamiliar vision, where objects near invisible to the naked eye were heightened through their tonal contrasts. As Hauser puts it, 'trees, fields, church towers, towns; the receding orders of the earth and sky, foreground, middle ground and misty distance – were all made unfamiliar from the air, all turned inside out'.[57] The angle of the aircraft, of the camera, and of the sun's rays could reveal much to the discerning and trained aerial eye.

The camoufleurs were to learn that effective camouflage demanded an intimate knowledge of vertical and oblique aerial photography and of the difference between them. In vertical aerial imagery, the camera is fixed in a vertical mounting, with the lens pointing directly downwards through a hole in the fuselage or via a side-mounted position. The vertical photograph thus presents a view of an area from a point immediately above it (Figure 10.2). In oblique aerial photography, the camera is fixed (or hand-held) so that it points towards the ground to the horizon, at an angle normally between 15 to 40 degrees (Figure 10.3). The oblique thus represents a perspective view of the area pictured. It is more difficult in oblique photography to compare the area covered with the corresponding area on the map.[58] In general, the vertical photograph is more valuable for military purposes than the oblique. The technique of operation was simple and the resulting image

---

54   Imperial War Museum [hereafter IWM], DRA 220/01-04, CD&TC Film *Air View* 1941.

55   DG GMA, A64/1/16/2/2, Booklet: 'Backgrounds' [1942].

56   IMW 86/50/37, Lectures and Training notes.

57   Kitty Hauser, *Shadow Sites: Photography, Archaeology, and the British Landscape* (Oxford: Oxford University Press, 2007), 172.

58   IWM 91/2/1, Instructor's Lecture notes 'Photographic Interpretation'.

**Figure 10.2   An example of the vertical angle in aerial views,
from British aerial photography taken in May 1940**

Source: By permission of the Imperial War Museum. © Crown Copyright IWM.

could be related to maps of the area because linear and angular measurements could be made, permitting the photographs to be more accurately interpreted and incorporated into a wider geographical and military knowledge of the battlefield.[59]

These differences did not, however, immediately provide a distinct solution for the camoufleurs. In all probability, the vertical image would be the angle that an aerial reconnaissance photographer would fly over an enemy's territory, but the bomber's view was from an oblique angle.[60] The camoufleurs were, thus, required to become closely acquainted with the geography and geometry of bombing from the view point of the bomber. They learnt that a bomb does not

---

59   Ibid.
60   IWM 86/50/3/2u, Camouflage Lecture Notes.

**Figure 10.3   An example of oblique angle aerial photography,
here over Norfolk in February 1942**
Source: By permission of the Imperial War Museum. © Crown Copyright IWM.

fall vertically; its flight path traces a parabolic curve as it acquires and then loses
some of the forward momentum of the plane. Because this is so, bombs must leave
the aeroplane while it is still some distance from its target, and so, crucially, the
bomber must identify that target from a range of several miles.[61] Camoufleurs had
to respond to this challenge.

Designing Disruption

As the camoufleurs were to discover, dialogue between these two angles and the
potential for disruption between the two aerial perspectives of the camera (vertical)
and bomber (oblique) was a requirement. The camoufleurs had to make potential
military targets blend or appear indistinct from the aerial perspective in order so
the bomber would be confused, and so might hesitate even if only for an instant,
over the exact location of the target. Although it was virtually impossible for

---

61   Ibid.

camoufleurs to conceal perfectly a military presence or to hide military installations entirely from view, they could design to create doubt. The camoufleurs realised that they could not confuse the glass lens of the camera. But they could attempt to confuse the fallible human eye of the aerial photographer who, moving at speed, operated the camera. In their training handbook the camoufleurs were told:

> It is one thing for the air photograph interpreter, sitting at ease with his feet up and a glass of beer by his side to discover all about you, but the bomber who comes primed with the information he has elucidated will not be in the same happy position. He will only have seconds to pick out on the ground that which may take quite a long time to discover on the photographs.[62]

The human eye of the bomber pilot works less systematically then the technological lens of the camera, and at a less leisurely pace then of the aerial interpreter. Accordingly, camoufleurs set out to invent ways to create doubt and indecision. To achieve this end, they became masters in trickery and deception, both by concealing and disrupting the background pattern of the earth, and by anticipating the difference between the 'seeing eye' of the aerial photographer and the bomber. Camouflage was developed to exploit the fallible human optic through understanding visual practices and becoming acquainted with the angular visuality of aerial space. Camouflage thus became a craft of erasure developed in symbiotic relationship with aerial technologies as, in turn, the roles of observer and observed became interchanged and blurred.

By studying how the camoufleurs visualised and interacted with the battlefield, it becomes apparent that to them, space was a cubic medium with which to work; airspace created and enabled a continual dialogue between earth and sky, from and through vertical and oblique angles. The camoufleurs' appreciation of the geometry of bombing and their efforts to subvert it reveals that aerial space is cut through with a complex criss-cross of angles, planes, trails, and shadows. Airspace is busy with the contingent and constantly shifting. Although not separate from the spatial experiences of the reconnaissance pilot, the photographic interpreter and the bomber, what made the camoufleurs' relationship with space distinct was the continuous engagement of eye and body that camouflage technology demanded. If, as has been claimed, the aeroplane developed new ways of understanding time and space, the development of camouflage technology in World War II shows how those new modes were also fallible and could be undermined.[63]

---

62   DG GMA, A64/1/16/2/4, Booklet 'Camouflage' [Spring 1942].

63   Peter Adey, *Aerial Life: Spaces, Mobilities, Affects* (Oxford: Wiley-Blackwell, 2010).

Conclusion

Camoufleurs sought to ground the aerial view through close study by understanding the aerial view as an angular geography of textured landscapes, while, at the same time, aiming to make elements of that terrestrial world invisible through concealment. Geographer John Wylie has argued that 'perhaps all landscapes, rural and urban, artistic and topographic, could be examined in terms of the tensions they set up and conduct between observer and observed, tensions between ways of seeing and interacting'.[64] In examining this tension between the observer and observed in landscape, he suggests that 'sometimes the observer/observed couplet might be intimate and tactile'.[65] For the camoufleurs, their battlefield was visualised and, if embodied, was so at a distance. Yet their relationship with the battlefield was as observer and as observed. The camoufleurs developed technology by developing concealing netting and scrimming in suitable colours and materials, by use of blending and appropriate paint shades and by employing designs that incorporated dazzle and coincident patterning, along with more evidently deceptive techniques such as the staging of decoys and dummies.[66] This was a technological regime of practice that responded to the demands of modern warfare because it interpreted terrain as the interconnection of tones, textures, technologies, and elements. In their engagement with space, camoufleurs became the masters of concealment; they were by some even described as the handmaidens of deception.[67]

In Britain the need for such protective camouflage was first evident by the late summer of 1940 as the country, specifically its cities, came under aerial attack. Because Britain was prepared and won what became known as 'The Battle of Britain', German bombing switched from the precision bombing of airfields and military installations to area bombing. This involved mass raids 'usually by hundreds of heavy bombers arriving in waves'.[68] Civilian morale in turn became an object, something 'known, rendered actionable, and intervened on'.[69] This shift in bombing strategy is important in understanding the history of camouflage. As Ken Hewitt has identified, area bombing was only seriously employed after the 'inability of bomber crews to make precision raids against well-defended or distant targets'.[70] Camouflage anticipated the fallible human optic and engineered moments of confusion, leading the bomber pilot to hesitate, to mistrust his eyes, the map, and the carefully-interpreted aerial intelligence.

---

64   John Wylie, *Landscape* (London: Routledge, 2007), 9.
65   Ibid., 9.
66   Trevelyan, 'Camouflage', 69.
67   Wiseman, *Visual and Sonic Warfare*, 4.
68   Hewitt, 'Place of Annihilation', 261.
69   Ben Anderson, 'Morale and the Affective Geographies of the "War on Terror"', *Cultural Geographies* 17 (2010): 219–36, quote from page 223.
70   Hewitt, 'Place of Annihilation', 261.

Rachel Woodward has called for military geographies to 'make war real' and to 'bring the battles back to the home front'.[71] Through its promise of protection and its basis in visual deceit, camouflage was an indication of the proximity of war and destruction for the British population, and it was an attempt to guard against death from distance. In World War II, camouflaged guns, factories, and defences individually and severally presented a starkly visible signal that the enemy was watching and that danger was an ever present. But, while it may not have been of itself a form of offence, camouflage was not benign. It was developed in order to protect important military positions, infrastructure and materials, not to hide civilian bodies. This chapter has demonstrated that developments in camouflage technology in World War II were directly connected to advancements in the technologies of surveillance and targeting. The development and adoption of techniques of deception in order to outwit the ever mobile technologies of warfare marked a shift in the British military's willingness to adopt covert and obscure methods of conflict. The development of camouflage as a subtle and subverting instrument of war required the military to draw in civilians from the worlds of art, science, and theatre. Britain's war-time camoufleurs studied each other's craft, aerial visualities, techniques of observation, and airborne targeting in order to produce a technology that was designed to go unnoticed. By exploring the establishment and training of the first recruits to the CD&TC, this chapter has revealed camouflage to be a tactical and tactile technology, its form and application being rooted in a complex relationship between predator and prey, the scientist and the instrument, the artist and visual literacy, the military and the civilian, aerial perspectives and the grounded understanding of terrestrial warfare. By developing and extending a technology that could subvert and undermine the aerial perspective while also, seemingly, playing tricks with terrestrial geography, camoufleurs designed a deceptive instrument which had lasting consequences for knowledge and technology and for the strategies and ethics of modern warfare.

---

71   Woodward, 'From Military Geography', 732.

Chapter 11

Instruments of Science and War:
Frank Malina and the Object of Rocketry

Fraser MacDonald

Some feats of exploration, certain forms of observant travel, seem to get all the attention. The voyages of Christopher Columbus, James Cook's Pacific travels, or the Apollo Lunar landings have each, in different ways, come to dominate the very idea of exploration. Arguably, some voyages are significant less for their discovery of new territories or spaces than for what they do to 'open up' the technologies and instruments of exploration; as a result, the achievements of the individual are sometimes harder to mobilise for the reputational gain of the state. This chapter follows Frank J. Malina, one of the most important, and one of the most neglected, twentieth-century engineers whose technology helped inaugurate the Space Age but whose success has seldom been championed by the state he served. Malina was one of the founders of modern astronautics and his legacies in the field of jet propulsion – both theoretical and institutional – were central to the success of rocketry as a vehicle of scientific exploration and instrumental measurement. The technology he initiated, in turn, propelled the development of atmospheric and solar physics, ultraviolet and X-ray astronomy as well as microgravity research. Malina gave shape to the rocket as the primary vehicle for what we now call the 'space sciences'. His work helped create the means by which humans for the first time in history were able to cast their instruments beyond Earth's atmosphere.

In the historiography of spaceflight, however, Frank Malina remains a relatively marginal figure. The conventional history of rocketry usually proceeds via a series of canonical developments and biographies. These include: twelfth-century Chinese 'arrows of fire'; Sir William Congreve's cast-iron munitions for the Napoleonic Wars; the influential descriptions of space travel by Jules Vernes; the early mathematics of spaceflight sketched from the primitive conditions of Konstantin Tsiolkovsky's Russian log cabin; the publication in 1923 of Hermann Oberth's *By Rocket into Planetary Space* which became one of the *ur*-texts of practical rocketry; American physics professor Robert Goddard's 1920 paper 'A Method of Reaching Extreme Altitudes', unfairly ridiculed by *The New York Times*;[1] early experiments with liquid rocket fuel by French aviator Robert Esnault-Pelterie; and Sergei Korolev's, formative role, shrouded in Politburo secrecy, in

---

1 Robert H. Goddard, "A Method of Reaching Extreme Altitudes", *Nature* 127 (1920): 809–11.

the development of the Soviet space programme. By far the outstanding figure, however, is Wernher von Braun, the theoretical and logistical genius behind the V-2 rocket – the man, as satirist Tom Lehrer once sang, 'whose allegiance was ruled by expedience'. Brushing aside his SS-membership and his proximity to war crimes, von Braun not only took America to the moon but also charmed his way on to the televisions of a grateful nation. A new wave of critical historiography concerning the V-2 programme has brought out the continuities between the engineering of Nazi terror and American transcendence.[2] Unsurprisingly perhaps, given von Braun's contribution to the Apollo programme, the history of American rocketry remains orientated towards the primacy of the V-2.

Frank Winter, one of the most venerable space historians, once insisted that 'it is generally recognised that all modern launch vehicles have derived from the V-2'. Such an exaggeration might have been unremarkable were it not made in a personal letter to an unimpressed Frank Malina.[3] Winter's *Smithsonian Guide to Spaceflight* ('the complete illustrated history from earliest designs to plans for the 21st century') makes no mention of Malina despite its publication by the very institution tasked with preserving national achievements in air and space. Walter A. McDougall's Pulitzer Prize-winning *The Heavens and the Earth: A political history of the Space Age* is one of the most complete works on the history of space exploration and yet, across 500 pages, it has more index entries for Bob Dylan than for Frank Malina.[4] While there is a small but important body of scholarship on Malina, it has had limited success in securing his public reputation as one of the founders of modern astronautics.[5]

On what grounds then might Malina be so regarded? Malina led the team that pioneered the world's first sounding rocket, the WAC Corporal. This was America's first high altitude rocket, a vehicle could that travel higher than balloon technology of the time. It is thus with Malina and his team, rather than Robert Goddard, that we can find the first *successful* liquid propellant spaceflight from the non-German branch of rocketry. Malina together with his Caltech contemporaries Tsien Hsue-shen and Martin Summerfield among others, also sketched out much of the early theory and application of jet propulsion. Malina's 1947 paper, co-authored by Summerfield, laid out for the first time the theoretical criteria for

---

2   See in particular the outstanding work of Neufeld: Michael Neufeld, *The Rocket and the Reich: Peenemünde and the Coming of the Ballistic Missile Era* (Boston: Harvard University Press, 1996); Michael Neufeld, *Von Braun: Dreamer of Space, Engineer of War* (New York: Alfred A. Knopf, 2007).

3   Frank J. Malina Collection, Caltech Archives [hereafter FJM Collection], Letter from Frank H. Winter to Malina, 24 March 1978, Box 6, Folder 31.

4   Walter A. McDougall, *The Heavens and the Earth: A Political History of the Space Age* (New York: Basic Books, 1985).

5   For instance, Clayton Koppes, *JPL and the American Space Program: A History of the Jet Propulsion Laboratory* (New Haven: Yale University Press, 1982).

'staged rockets' – the standard means of orbital access for over half a century.[6] With his colleague Jack Parsons, Malina developed an entirely new category of castable solid rocket fuel, GALCIT-53, that would become a primary precursor of America's Space Shuttle boosters as well as of numerous ballistic missiles. They also pioneered hypergolic propellants – where fuel and oxidiser spontaneously ignite – that would later be used in the Apollo programme. Though the chemical inspiration came from Parsons, Malina and his mentor Theodore von Kármán worked out the theory behind long-duration restricted-burning solid fuel.[7]

There's another reason too: when the WAC Corporal was 'mated' on to the V-2 – the so-called BUMPER WAC – it became the first high altitude staged rocket. And on 24 February 1949, the flight of BUMPER no. 5 reached an altitude of 244 miles, becoming the first human object to reach into what was then defined as extra-terrestrial space. It is this largely unsung achievement, eight years before Sputnik, that arguably marks the beginning of the Space Age. The flight may have lasted just 390 seconds but it is a voyage as remarkable in its own way as those of Columbus, Magellan, and Cook. Malina's WAC Corporal was also the precursor to the US Corporal missile – the first missile to be authorized to carry a nuclear warhead and the progenitor of contemporary weapons of mass destruction. Lastly, Malina together with von Kármán founded two of the enduring institutions of space exploration: the Jet Propulsion Laboratory (JPL), now a NASA research centre with a $1.5 billion annual budget best known for its exploration of the Martian surface with its *Curiosity* rover; and Aerojet, one of the key rocket propulsion manufacturers of the twentieth century – still trading today as Aerojet Rocketdyne.[8]

Malina's work deserves fuller treatment that can be given here. In this chapter I examine his contributions to astronautics in order to address two themes: about the reciprocity between instrument and vehicle; and about the co-dependence of scientific and military rationales in the development of rocketry. In the first instance, I draw upon this neglected history of rocketry to argue that the *practices* and *devices* of instrumental measurement are not always easy to distinguish from the *vehicles* for instrumental measurement. In one sense this is not a new argument. Richard Sorrenson makes the case in his 1996 essay on eighteenth-century instruments at sea that the ship 'was never merely a vehicle', and that the particularities of the British naval ships used by Cook, among others, gave shape to the voyages of exploration.[9] The ship, for Sorrenson, is itself a kind of instrument,

---

6   Frank J. Malina and Martin Summerfield, 'The Problem of Escape from the Earth by Rocket', *Journal of Aeronautical Sciences* 14 (1947): 471–80.

7   Theodore von Kármán and F.J. Malina, 'Characteristics of the Ideal Solid Propellant Rocket Motor' in *Collected Works of Theodore von Kármán*, Vol. IV, (London: Butterworth, 1956), 64.

8   Frank H. Winter and George S. James, 'Highlights of 50 Years of Aerojet, a Pioneering American Rocket Company, 1942–1992', *Acta Astronautica* 35 (1995): 677–98.

9   Richard Sorrenson, 'The Ship as a Scientific Instrument in the Eighteenth Century', *Osiris* 11 (1996), 221–36, quote from page 222.

the observations from which were a function of its design, operating procedures, and onboard discipline. He goes on to explore a suggestive correspondence between the ship in the mid-eighteenth century and the 'rocketship' of the mid-twentieth century. Drawing on the work of David DeVorkin, Sorrenson writes that the V-2, as a vehicle or tool, 'could be considered an instrument that, like a ship on a voyage of scientific discovery, leaves a trace of its interaction with the medium it passes through'.[10] It is this suggestive idea that I aim, in part, to develop here – albeit in relation to America's home-grown rocketry programme that the V-2 so successfully eclipsed.

My argument is also a little different from that of Sorrenson. It is not just that vehicles can be thought of as instruments but rather that instruments and vehicles alike are co-produced within a recursive and developing field of measurement. To put it another way, the rocket is the object of repeated instrumental scrutiny in order that it should be remade anew as a more refined and predictable bearer of yet further forms of instrumentation. In this instance, vehicle and instruments are practically indivisible. The rocket is, in other words, the epistemological and technological outcome of a series of attunements – some successful, others less so – based on the precision measurement of performance characteristics. The principal objects of measurement here are thrust, power, and efficiency. The term 'specific impulse' would eventually be used as a measure of the force generated by a given amount of propellant in a unit of time. This characteristic, however, was yet to be defined; it is the very development of such objects of measurement, as well as the instruments of measurement, that constitute the multivariate puzzle of building a viable rocket motor in the first place.

A second theme of the chapter considers the entwining of scientific exploration and military power as dual objects of rocketry from the late 1930s.[11] The idea of the rocket as a vehicle for upper atmosphere research was the primary impetus behind jet propulsion research at Caltech. But the anti-fascist politics of the engineers, combined with their growing financial reliance on the US military, meant that a technology designed to carry instruments was, at the same time, intended to bear deadlier payloads. As World War turned to Cold War, the two defining military technologies of the twentieth century – the atom bomb and the rocket – would be fused in the totem of the nuclear missile. Making space for instruments was thus simultaneously mixed up with other technologies of defence and conquest within

---

10   David H. DeVorkin, *Science With A Vengeance: How the Military Created the US Space Sciences After World War II* (New York: Springer, 1993).

11   The co-production of scientific and military knowledge in the aftermath of World War Two is now well established: see, for example, Simone Turchetti and Peder Roberts, 'Introduction: Knowing the Enemy, Knowing the Earth' in *The Surveillance Imperative: Geosciences during the Cold War and Beyond* (ed.) S. Turchetti and P. Roberts (London: Palgrave, 2014), 1–19; Ronald E. Doel, 'Constituting the Postwar Earth Sciences: The Military's Influence on the Environmental Sciences in the USA after 1945', *Social Studies of Science* 33 (2003), 635–66; DeVorkin, *Science with a Vengeance.*

the frame of Westphalian territoriality. And it is in the creation of the modern rocket that we can witness the historically significant shift from Sorrenson's union of instrument and vehicle to the fuller conjunction of instrument-vehicle-projectile.[12]

These two themes are here examined with regard to different episodes in the making of America's JPL rocket programme. In three sections I outline some of the instrumental procedures that made the rocket, as well as those for which the rocket was made. The first details the conception and development of a viable rocket motor, a process in which engineers creatively improvised with materials, instruments and funding, to successfully measure thrust. This is, I argue, a case of rocketry as the episto-technological outcome of precision measurement. The second section examines the culmination of Malina's earlier successes in the engineering of thrust to enlist military support to build a research vehicle, a sounding rocket, to study the upper atmosphere. This is the rocket in the guise of an instrument carrier, the WAC Corporal being the first such vessel designed to deliver and retrieve the instruments of space science. The third section considers the rocket as itself an instrument: calibrated to produce, and eventually to transmit data about its trajectory – traces, as Sorrenson has it, of 'its interaction with the medium it passes through'.[13]

Nativity: Experiments with a Rocket Motor

The earliest evidence of Frank Malina's research interest in rocketry can be found in letters sent to his parents in small town Texas. As a young masters student at GALCIT – the Guggenheim Aeronautical Laboratory at the California Institute of Technology – he diligently kept them up to date with news of his studies (aeronautical engineering), paid work (as an assistant on the Caltech wind-tunnel), dating (too busy), reading (the Soviet-friendly *The New Republic* magazine) and his new circle of friends. 'Yesterday I resumed my lessons in Russian' he told them, noting that his teacher, 'Dr Weinbaum, has invited … me to his house for Tuesday evening'.[14] Sidney Weinbaum was a brilliant Russian emigrant: a prize-winning chess player, a concert-standard pianist and a research assistant – a 'human computer' – to the famed Caltech chemist Linus Pauling.[15] The close friendship between Malina and Weinbaum, and their habit of meeting in the evenings with others to discuss music, art and politics, would be subject to much scrutiny in

---

12   Paul Virilio, *Bunker Archaeology* (Princeton: Princeton Architectural Press, 1994), 20.

13   Sorrenson, 'The Ship as a Scientific Instrument in the Eighteenth Century', 222.

14   Letter from Malina, 2 February 1936, Box 21, Folder 6, FJM Collection.

15   Thomas Hager, *Force of Nature: The Life of Linus Pauling* (New York: Simon and Schuster, 1995), 660; Mary Jo Nye, 'Mine, Thine, and Ours: Collaboration and Co-Authorship in the Material Culture of the Mid-Twentieth Century Chemical Laboratory', *Ambix* 61 (2014), 211–35.

the McCarthy era. But the same letter on 2 February 1936 concluded with the observation that: 'am looking for a new subject to start working on' ... 'wouldn't be surprised if I got my fingers mixed into rocket propulsion'.

The following week's letter brought news of an interesting gathering at the Weinbaums as well as notice of Malina's intention to approach Clarke B. Millikan, a professor of aeronautics, with the idea of doing rocket research. By 14 February 1936, Malina's parents learned that 'two fellows from Pasadena have interested a student named Bollay and me in making a rocket motor with which to conduct a series of tests'.[16] William Bollay is not a name that has any lasting significance in the annals of rocketry but the same may not be said of the 'two fellows'. John Whiteside Parsons – Jack Parsons, as he was universally known – was an occultist and self-taught chemist whose education came from working in an explosives factory. His schoolfriend Ed Forman was a skilled mechanic and machinist. Ever since boyhood they had blackened their lawns with rocket experiments before recognising that a more rigorous scientific approach might more plausibly propel their ambitions.

It was an unlikely if productive partnership. The elective affinity between Malina and Parsons was obvious and complimentary; where Malina prized the rigours of mathematical theory, Parsons preferred hands-on experimentation. Added to this triad would come other Caltech students – most of them, like Malina, supervised by the great aerodynamicist Theodore von Kármán – who shared their curiosity about this nascent field. There was the Chinese theoretician Tsien Hsue-shen; the aeronautical engineer Apollo Milton Olin Smith; and, later, the physicist Martin Summerfield. Even at the outset, when the rocket was scarcely a sketch on paper, the purported object of such a trial was clear in Malina's letter home: 'If we could develop a rocket to go up and come down again safely we would be able to get much data useful to the weather men and also for cosmic ray study'.[17]

Preparing for a static test

A year after meeting Parsons and Forman, Malina told his parents that:

> plans for making experiments with a rocket motor are getting well along. Next week parts will be cast ... First we want to make a detailed study of the power available and then build a rocket to go to altitude above those reached by balloons for meteorological purposes. Very little is known of conditions of existing at 100,000 feet and above.[18]

The need to experiment with a static test – to measure the power output of a rocket motor before attempting to leave the ground – was the hallmark of any von

---

16   Letter from Malina, 14 February 1936, Box 21, Folder 6 FJM Collection.
17   Ibid.
18   Letter from Malina, 21 February 1936, Box 21, Folder 6, FJM Collection.

Kármán disciple. 'Until one could design a workable engine' wrote Malina, 'there was no point in devoting effort to the design of the rocket shell, propellant supply, stabilizer, launching method, payload parachute etc.'.[19] But these static tests were both preceded and succeeded by rigorous theoretical studies of the reaction principle and of the flight performance requirements of a sounding rocket. The reaction principle is enshrined in Newton's third law of motion: for every action there is an equal but opposite reaction. At GALCIT this was well understood in relation to propeller dynamics but not in relation to rocketry.

As head of GALCIT, von Kármán had only a minimal acquaintance with theoretical rocketry; he had been preoccupied with turbulence theory and supersonic flow problems in aircraft. But notwithstanding rocketry's prior association as a hobbyist enterprise he could see potential in this eccentric new field. By his own admission, he was 'immediately captivated by the earnestness and the enthusiasm of these young men'. They 'set forth modest aims' he recalled: 'they wished to build and test liquid and solid rockets which could be propelled perhaps twenty to fifty miles into space'.[20] The ultimate object was the same for von Kármán as for Malina: 'a small rocket when properly instrumented could bring back information about cosmic rays and weather at the edge of outer space'. Von Kármán gave the group his blessing and also agreed to supervise Malina's PhD thesis on the performance characteristics of the rocket motor.

Part of the difficulty faced by the Rocket Research Project, as it was then known, was in sourcing equipment and materials with which to make their tests. By June 1936, Malina wrote home to say that 'Parsons and I drove all over L.A. looking for high pressure tanks and meters. Didn't have any luck. Two instruments we need cost $60 a piece and we are trying to find them second hand. I am convinced it is a hopeless task'.[21] Progress in the research depended on their willingness to improvise with equipment and funding alike. The most improbable solution to the finance problem was for Parsons and Malina to write a novel about socialist rocketeers who have to keep their designs from falling into the hands of corporate bosses and Nazis; this plot in turn could be offered to Hollywood as a movie script. 'The story is to be built on the present stage of rocketry as the foundation', opened the précis 'with a superstructure of the dynamic social problems now existing'.[22] It is not known whether MGM, the studio which had that summer snapped up the rights to *Gone with the Wind*, had the chance to bid on this incombustible mix of rocketry and Marxism.

---

19 Frank J. Malina, 'On the GALCIT Rocket Research Project, 1936–1938', *Smithsonian Annals of Flight* 10 (1974), 113–27, quote from page 117.

20 Theodore von Kármán, *The Wind and Beyond: Pioneer in Aviation and Pathfinder in Space* (Little, Brown: New York, 1967), 238.

21 Letter from Malina, 29 February 1936, Box 21, Folder 6, FJM Collection.

22 'Summary of Preface', MGM Project, unpublished manuscript, Frank J. Malina private collection, Malina Family Archive [hereafter MFA].

Making and Measuring Thrust

Eight months after Malina indicated a research interest in rocketry, he and his collaborators were finally ready to attempt a static test. Little of direct benefit had been learned from an exchange with Robert Goddard who, burnt by *The New York Times* ridicule, had proved secretive about his own designs. Though Goddard had driven Malina out to his desert test and launching tower facility in Roswell, New Mexico, the visit had proved of little use to the Caltech team. As von Kármán would later put it 'there is no direct line from Goddard to present day rocketry. He is on a branch that died'.[23] Malina's static test set-up was a more improvised and impermanent affair in comparison to that of Goddard. Everything had to be carted from GALCIT to an out-of-the-way site on the dry river bed of the Arroyo Seco, not far from the Devil's Gate Dam – a laborious process to be fitted around classes and time doing paid work on the wind tunnel. On the morning of Saturday 31 October 1936, the nativity scene of American rocketry, the Rocket Research Project conducted their first test using a small rocket motor cooled with a primitive jacket of water, using gaseous oxygen as an oxidant and methyl alchohol as fuel (Figure 11.1).

Success came initially in the guise of failure. 'Very many things happened that will teach us what to do next time', wrote Malina to his parents. 'The most excitement took place on the last "shot" when the oxygen hose swung for some reason ignited and swung around on the ground, 40 ft from us. We all tore out across the country wondering if our check valves would work'.[24] Thrust was to be measured by an improvised diamond scratching a trace on a glass cylinder but as none of their four tests ignited, the glass remained pristine. By January 1937, however, the same set-up saw the motor run uninterrupted for 44 seconds at a chamber pressure of 75 lbs per square inch. Given the hopeful results of these initial static tests, von Kármán agreed to allow the rocketeers use of GALCIT facilities to conduct smaller scale tests – saving them the constant burden of assembling and dismantling equipment in the Arroyo Seco. The theoretical analysis was in time written up by Malina and Smith for the *Journal of the Aeronautical Sciences*: 'if a rocket motor of high efficiency can be constructed' concluded the authors, 'far greater altitudes can be reached than is possible by any other known means'.[25] A seminar given by Malina in April 1937 finally catalysed a solution to their funding problem when a meteorology student, Weld Arnold, was sufficiently excited by rocketry's potential to offer the team $1,000, a benefaction of uncertain provenance that has since become part of JPL's myth of origin. This unexpected gift not only solved a practical problem – how to pay for materials in the absence of literary success – it also reflected the fact that the project had caught the interest

---

23 Von Kármán, *The Wind and Beyond*, 242.

24 Letter from Malina to parents, 1 November 1936, Box 21 Folder 6, FJM Collection.

25 Frank J. Malina and A.M.O. Smith, 'Flight Analysis of a Sounding Rocket', *Journal of the Aeronautical Sciences* 5 (1938), 199–202.

Figure 11.1   Malina and Parsons at their first static test
Source: Courtesy of the Malina family.

and moral support of other Caltech scientists who were excited by its application as a vehicle for instrumental measurement. Foremost among those who saluted their achievements were Caltech President and Nobel Prize-winning cosmic ray pioneer Robert A. Millikan (father of the initially circumspect Clarke B. Millikan), as well as faculty meteorologist Irving Krick.[26]

For all this support, the rocketeers would learn that Caltech's scientific goodwill had its limits. Campus rocketry experiments had the potential to disrupt as well as to advance science. A smaller scale test – this time with nitrogen dioxide as oxidant – was conducted by mounting a motor with propellant on the bob of a 50 foot pendulum suspended from the third floor of the GALCIT laboratory. In this case, thrust was ascertained by measuring the simple deflection of the pendulum. But when the motor misfired, the problem of constant combustion being at that time unsolved, a noxious mist of NO_2-alcohol permeated the building leaving a fine layer of rust on most of the laboratory's permanent equipment.[27] The team were unceremoniously evicted though not before being presented with oily rags to undo some of the damage. This incident had its losses and gains. A significant loss was that of the lab space that had been hard-earned by the success of the first Arroyo Seco tests. A dubious gain was a new name by which they were now known: 'the suicide squad'. More importantly, they had for the first time experimented with a storable liquid oxidizer – a development that would anticipate the eventual use of Red Fuming Nitric Acid (RFNA) as the oxidant in the Corporal missile.[28] They had also improvised with the instrumental measurement of thrust. Scholer Bangs, a minor but resonantly named historian of astronautics, recalled meeting Malina shortly after this eviction from the GALCIT lab:

> On a concrete walk outside the Aeronautics building Malina, getting closer to his PhD, had set up a small test stand and clamped on it one of his small "gunpowder" rockets. To measure thrust he used a spring scale once used in a grocery store. Malina asked spectators to step back. He closed a dime-store electric switch. The rocket went "bang!". The needle on the grocery scale jiggled. That was the conquest of space.[29]

The early exploration of space is thus to be found in this series of terrestrial experiments in the production and measurement of thrust. Though there are many earlier tests – for instance those of Robert Esnault-Pelterie in France, Eugen Sänger in Austria, Hermann Oberth and others in the German *Verein für Raumschiffahrt* (Spaceflight Society) – the Caltech work stands out as the primary branch of high-

---

26   Malina, "On the GALCIT Rocket Research Project, 1936–38", 115.

27   Frank J. Malina, "The Rocket Pioneers: Memoirs of the Infant Days of Rocketry and Caltech", *Engineering and Science* 31 (1968), 9–32, quote from page 30.

28   Malina, "On the GALCIT Rocket Research Project", 121.

29   Untitled memoir manuscript by Scholer Bangs, 2.5.69, JPL, Box 5 Folder 7, FJM Collection.

altitude rocketry aside from the V-2.[30] As the world braced for war, the Caltech team remained driven by an interest in exploring the upper atmosphere while also remaining sympathetic with the war effort. Science was the object; the rocket was the vehicle. It was also a vehicle that did not yet exist: it was a distant and deferred goal rendered subservient to a theoretical understanding of the performance characteristics and combustion dynamics of a rocket motor. In this sense, vehicle and instruments were co-produced – founded on countless tiny measurements, a culture of engineering practice that would later be institutionalised in the shape of the Jet Propulsion Laboratory. A similar co-production might also be traced in the bespoke shipyard adaptations for eighteenth-century exploration discussed by Sorrenson. Unlike oceanic voyages, however, where the place of the craft was determined by the ship's captain and crew rather than by boatbuilders, the early exploration of space was achieved through instrumental procedures of design and production that are difficult to disentangle from those of launch and operation. These early traces of success – a scratch in glass, the jiggle of a grocery scale – may be more ephemeral than those imagined by Sorrenson but they were the registers on which viable spaceflight was first recorded.

Making Space for Instruments: The WAC Corporal Sounding Rocket

The name Jet Propulsion Laboratory was coined to circumvent a troublesome word: rocket. Though various branches of the US military would become interested in potential applications, no one in wartime was keen to invoke either the amateur connotations, or the hyperbolic sci-fi imaginations, of 'rocket propulsion'. So when the GALCIT Rocket Research Project eventually secured serious military funding for its work, the disavowal of the earlier history – that of amateur engineers – was almost part of the contract. At the end of 1938, von Kármán had sent Frank Malina to Washington to address the National Academy of Sciences Committee on Air Corps Research.[31] Duly impressed, the Army Air Corps commissioned new research on the use of rocketry as an auxiliary propulsion system for aircraft. The need for shorter take-off distances for bombers was a new priority: Jet-Assisted Take-Off units – JATOs – became the first application of the Rocket Research Project's work, the production of which fell to a new company, Aerojet, that the rocketeers had established with von Kármán. After the attack on Pearl Harbor and the recognition that the US air war in the Pacific would depend on small island runways, JATOs turned out to be big business. The real breakthrough had come from Jack Parsons who, encouraged by theoretical work by Malina and von

30 Frank H. Winter, *Prelude to the Space Age: The Rocket Societies, 1924–1940* (Smithsonian: Washington, 1983).

31 Frank J. Malina, 'The Origins and First Decade of the Jet Propulsion Laboratory', in *The History of Rocket Technology* (ed.) Eugene M. Emm (Wayne State University Press: Detroit, 1964), 46–66.

Kármán, had devised a castable solid restricted-burning propellant, GALCIT 53, using potassium perchlorate as oxidant and asphalt as fuel and binding agent.[32] Another programme for a liquid propellant JATO – using aniline and RFNA – saw similar success in 1941. But at this stage no part of the US military showed serious interest in a rocket motor propelling anything other than aircraft.

All that changed in 1943. When European intelligence reports intimated the existence of a German programme for weaponised rockets, the US Army Air Force quickly commissioned a study exploring the possibility of ballistic missiles using rocket engines that the Aerojet team had already developed. Rockets, it was becoming clear, would have a postwar military future. But what kind of military future? Different fiefdoms within the forces vied for control. Army Ordnance laid claim to rockets on the grounds that, like a bullet from a gun, they followed a ballistic trajectory. The Air Force pointed to the indisputable kinship that rockets shared with pilotless aircraft, both requiring guidance and aerodynamic control.[33] In the summer of 1944, Army Ordnance demonstrated their enthusiasm by offering up to $3 million dollars for the first year of a programme of 'research, investigation and engineering in connection with the development of a long range rocket missile'.[34] 'This kind of money', admitted Malina, 'threw us into a proper dither!'[35] For all his concern, the Ordnance project ('ORDCIT') that followed was a well-conceived programme of research. It specified a remote controlled missile that could carry an explosive payload of 1,000 lbs for over 150 miles with an accuracy of less than 2 per cent dispersion at target – far beyond anything that had been developed to date. It also extended the complexity of rocket design into areas scarcely touched: electronics, guidance systems and telemetry. The engineering challenge of creating workable components that would be small and light enough to fly, and robust enough to withstand the stress of supersonic speeds, would in time make 'rocket science' a byword for complexity.

In just a few years, Malina had gone from being a PhD student experimenting with friends on a dry river bed to leading a multi-million dollar government research project employing hundreds of people now located in permanent facilities exactly adjacent to the nativity scene. The stock of his start-up company, Aerojet, rose as the order book swelled. Malina, however, was unhappy with the direction of his life and work. He had wanted to contribute to the war effort; indeed after the success of JATOs he told his parents that 'we now have something that really works

---

32   J.D. Hunley, 'Minuteman and the Development of Solid Launch Technology', in *To Reach the High Frontier: A History of U.S. Launch Vehicles* (eds) Roger Launius and Dennis Jenkins (Lexington: University of Kentucky Press, 2002), 229–300, quote from page 230.

33   Frank J. Malina, 'America's First Long Range Missile Program: The ORDCIT Project of the Jet Propulsion Laboratory, 1943–1946: A Memoir', *Essays on the History of Rocketry and Astronautics: International Academy of Astronautics*, Vol. II (1977), 339–83, quote from page 343.

34   Ibid., 345.

35   Ibid.

and we should be able to help give the Fascists hell!'.[36] The fascists, meanwhile, were developing their own rockets. In 1944, when Malina visited London as a military attaché – ostensibly to study the state of British rocket design – he was able to get first-hand experience of incoming V-1s and V-2s, fragments of which could inform military intelligence about German engineering. It would prove a stimulating experience, less from technical insights in rocket design, than from extending his cultural and political horizons among left-leaning British scientists like Joseph Needham and Julian Huxley. Being in the midst of the civilian target zone of a rival rocket programme did little to enthuse Malina about a rocket as a weapon of war. That this seemed like an inevitable development caused him much anxiety: 'I could not be happy doing the work I am now doing with the feeling that I was helping to prepare for another war rather than trying to stop it', he once wrote to his wife Liljan.[37] On the flight home aboard a B-17 bomber, Malina had what he later called a 'brainstorm'.[38] Plans were already well advanced at JPL for testing the first rocket of the ORDCIT contract – a small solid propellant missile called Private. The idea was that progressively refined and ambitious rockets would be made in order of military rank, starting with Private and working up to Colonel ('the highest rank that works' as von Kármán once teased the Generals). Amid the successful development of solid and liquid propellant motors for military purposes, what Malina realized on the B-17 was that the original goal of 'the suicide squad' in 1936 – that of a sounding rocket for scientific exploration – was now within reach. A liquid propellant sounding rocket with solid boosters (a staged or stepped rocket) could, as Malina told Army Ordnance, be the perfect interim vehicle to trial launch procedures for the scheduled Corporal missile. Malina's suggestion also coincided with the Signal Corps' need for an inexpensive instrument carrier that could carry 25 lbs of meteorological instruments to 100,000 ft or more and retrieved safely by parachute. It would be called the WAC Corporal: technically a small Corporal 'Without Attitude Control' (that is, no inertial guidance system) though few could resist the sexist interpretation of the WAC as the Women's Auxiliary Corps or 'Corporal's little sister'.[39]

Malina was finally building a rocket for scientific exploration, a vehicle for the research ambitions of other scientists. It was also, unhappily for him, a side project to develop long-range missiles (another justification was that it could help develop anti-aircraft missiles).[40] On one of his trips to the new White Sands Proving

---

36   Letter from Malina to parents, 22 March 1942, Box 21, FJM Collection.

37   Letter to Liljan Malina, 4 December 1945, MFA.

38   Malina, 'America's First Long Range Missile Program', 352.

39   See for instance William Pickering, Oral History by Shirley Cohen, 22 and 29 April 2003, Caltech Archives, http://oralhistories.library.caltech.edu/, 12; it is true also of Malina, see letter to Eugene Emme, 1 March 1972, Box 5 Folder 22 FJM Collection.

40   See Benjamin Zibit, *The Guggenheim Aeronautics Laboratory at Caltech and the Creation of the Modern Rocket Motor*. Unpublished PhD thesis, City University of New York, 1999, 447.

Ground, not far from Goddard's Roswell ranch, Malina directed the C-47 Douglas aircraft to detour via the new Trinity site at Alamogordo where an old acquaintance Robert Oppenheimer had overseen the first atomic test (Oppenheimer's brother Frank Oppenheimer had, like Malina, been a regular visitor at Sidney Weinbaum's house). 'It was a very disturbing sight', noted Malina 'especially for us who were involved in the development of long range missiles'.[41] The dual rationales for rocketry, though obviously different, were now ineluctably aligned: scientific exploration and warfare, transcendence and mass destruction.

With the WAC Corporal, Malina and the JPL team had created a vehicle with a cavity. Justified as a space for the instruments of science, it was more readily a space for the instruments of war. In the first instance, however, it was filled with concrete. An essential aspect of space exploration was to closely monitor the performance of every element in the complex system of the rocket. Concrete was a suitably inert ballast so that the team could assess isolated performance characteristics of, say, the booster or the liquid motor against the theoretical modeling already undertaken by Malina and Tsien. This systems approach was reflected in the organization of the WAC Corporal team, with specialists in charge of the solid rocket booster, the sounding rocket itself, launcher and nose cone, field test procedures, external ballistics and photography. They had learned from Goddard's failure that team work and a specialist division of labour were essential.

The success of the first round of WAC Corporal firings at White Sands came just two months after Goddard's death. True to the systems approach he had eschewed, the JPL firings were incrementally built up: first the Tiny Tim booster, then the WAC vehicle, then a partial charge of propellant to check motor function and separation procedures, and eventually the nose cone parachute release. On 11 October 1945, round 5 of the WAC Corporal soared into the azure sky to an altitude of 235,000 feet – well over double the Signal Corps' stated requirements (see Figure 11.2). The voyage lasted 450 seconds. It was the United States' first genuinely high altitude rocket. What had been demonstrated beyond all doubt, less than 10 years after the Arroyo Seco tests, was that a rocket could be refined as a more or less predictable vehicle for space exploration as well, of course, as a bearer of distant destruction. 'The WAC was a great success' claimed Bill Pickering, then head of remote control at JPL, 'as it offered for the first time, the realistic role of a scientific instrument carrier in a simple, relatively cheap form'.[42] Strictly speaking, Pickering was wrong: the WAC Corporal did not prove to be a great instrument-carrier. But its direct successor, the Aerobee, using the same propulsion technology and again made by Aerojet, would in time become the workhorse of the space sciences. Under the direction of physicist James Van Allen, the Aerobee took over from the US-launched V-2 as the primary vehicle for upper atmosphere research with over 1,000 launches from 1947 to 1985. The

---

41   Malina, 'America's First Long Range Missile Program', 364.

42   Douglas J. Mudgway, *William H. Pickering: America's Deep Space Pioneer* (Washington: NASA History Division, 2008), 48.

Figure 11.2   Frank Malina with the WAC Corporal sounding rocket, White Sands, September 1945

Source: Courtesy NASA/JPL-Caltech, photo number 293–364.

WAC Corporal had thus paved the way in almost every aspect of American rocket design and, though a very much smaller rocket than the V-2, did so at a fraction of the cost, infrastructure and personnel of its German rival.

Instrument-Vehicle-Projectile

In the midst of the 1946 testing at White Sands Proving Ground, Malina and others in the JPL team found themselves in the unlikely position of working alongside von Braun and the other Germans – engineers who, only a year or so earlier, had overseen the V-2 production by thousands of conscripted and brutalised slaves.[43] Under the auspices of Project Paperclip, Von Braun and his team were welcomed into the heart of the American space establishment in order that they might apply V-2 engineering insight to the United States long-range missile programme. The stark unseemliness of this situation – Nazi architects of destruction as born-again Americans – was not lost on Malina. While the United States looked the other way when it came to the war records of von Braun and his Nazi commander Walter Dornberger, the FBI had a growing interest in Malina's anti-fascist political activities in the late 1930s. By the time the FBI raided Malina's home in 1946, the prospect of a future building nuclear missiles had already prompted thoughts of an alternative career in international scientific cooperation.[44] In the meantime, however, he still chaired the JPL bi-monthly meetings coordinating the development of the Corporal missile as part of the original ORDCIT contract.

The changing place of instruments in the long genesis of the Corporal missile – from static tests of a rocket motor in the Arroyo Seco, through JATOs to the WAC sounding rocket – is instructive. The rocket is not only founded *on* instrumental measurement, nor merely conceived as a vehicle *for* the instruments of science. With the development of the WAC Corporal, the rocket also itself *becomes* an instrumental device, measuring the characteristics of the unfamiliar environment through which it eventually moved at supersonic speed. The continuities between ship and rocketship are of course apparent in the etymology of 'sounding' – measuring the envelope of Earth just as, on the same axis, earlier voyages ascertained the depths of its ocean. Sounding, in this case, is a scientific practice constituted by the design, launch, flight and tracking of the instrument-vehicle. But the practice of sounding was every bit as improvised in the upper atmosphere as it had been at sea.[45] Methods that should have theoretically been more accurate were not always reliable, while reliable methods – those, at least, that could dependably yield some

---

43   Neufeld, *The Rocket and the Reich*.

44   James L. Johnson, 'Rockets and the Red Scare: Frank Malina and American Missile Development, 1936–1954', *Quest* 19 (2012), 30–36, quote from page 31.

45   See for instance Sarah Louise Millar, 'Science at Sea: Soundings and Instrumental Knowledge in British Polar Expedition Narratives, 1818–1848', *Journal of Historical Geography* 42 (2013), 77–87.

kind of measurement – were not always accurate. With onboard altimetry not yet viable, it was radar tracking (then still in its infancy) that suggested a height, on 11 October 1945, of about 235,000ft. For most of the initial WAC Corporal flights, however, the radar failed completely. In the absence of radar, a visual analysis of the exhaust trail could be an important measure of progress, though only up to the height at which the motor stopped discharging propellants – around 80,000ft. This would be closely observed and photographed by three cameras with particular attention given to the 'smoke trace ... getting heavier or other peculiarities'.[46] A spectacular alternative in which combustion could be more visible was to stage a night launch as occurred on 25 October 1945.[47]

Not all relevant flight information could be determined at the time by observers on the ground: there were still important aspects of the journey that had to be visualized for analysis after the recovery of the rocket body. In at least one instance, a temperature sensitive paint was applied to parts of the WAC Corporal to disclose something of the relation between a supersonic body and its environment.[48] This meant, of course, that built into the design of the instrument-vehicle was a mechanism that would allow for its recovery in the first place. The primary purpose of the first WAC test firings had been to check the method of launch operation but the secondary purpose was to perfect the deployment of two parachutes – one to lower the rocket and the other to safely return the radiosonde and other instruments.[49] Not only did the radiosonde malfunction but, after a parallel failure of the parachutes, it plunged to a shallow desert grave.[50] Such was the difficulty of ensuring the smooth passage of a scientific payload.

Malina detailed the problem in a letter home to his wife Liljan. The explosive pins that tied the WAC's nose cone to its body had not functioned as expected (Figure 11.3). But he was not able to say this. Working on a classified military programme, he had grown used to writing around the security restrictions. 'Our "little lady" performed beautifully on her maiden flight' he wrote. 'We might be able to get her to do one more stunt that is expected of her before our program is completed. The habit is curious – she refuses to blow her nose when so instructed'.[51] But within a year this setback had led to a successful redesign to accommodate the difficulties of opening a parachute under the extreme low pressure of high altitudes.

---

46   William H. Pickering, 'Proceedings of Upper Atmosphere Symposium', in *Guided Missiles and Upper Atmosphere Symposium* March 13, 1946, Box 7 Folder 2, FJM Collection.

47   Malina, 'America's First Long Range Missile Program', 365.

48   Letter from Malina to George Flynn, Publisher of *Model Rocketry* magazine [undated but July 1970], Box 6 Folder 7 FJM Collection.

49   Frank J. Malina, 'Rockets for Upper Atmosphere Research', in *Guided Missiles and Upper Atmosphere Symposium* March 13, 1946, Box 7 Folder 2, FJM Collection.

50   James W. Bragg, *Historical Monograph No. 4: Development of the Corporal: The Embryo of the Army Missile Program* (declassified), Army Ballistic Missile Agency, xiii.

51   Letter to Liljan Malina, 11 October 1945, Malina Family Archive.

**Figure 11.3   Round 7, WAC Corporal parachute and nose cone test,
26 May 1946**

Source: Courtesy NASA/JPL-Caltech, photo number 293–433B.

Similar progress was made in the transmission of instrumental data by telemetry. In two of the early test firings of the WAC, the 10 channel FM-FM telemetry set successfully transmitted basic flight parameters to a ground station.[52] This too was a significant technological threshold: in 2014, JPL can still communicate with its Voyager 1 space probe over a distance of 12 billion miles.

All of these developments, in propulsion and in communication, would have made the WAC Corporal the primary vehicle for the space sciences were it not for the V-2. But it was the German rocket – built from 300 railroad cars worth of captured parts – that, from April 1946, became the vehicle of choice for the Upper Atmosphere Research Panel.[53] Under a contract to General Electric, 73 adapted V-2s were fired as part of the Hermes programme: nearly 50 per cent were

52   Mudgway, *William H. Pickering,* 52.
53   DeVorkin, *Science with a Vengeance.*

failures.[54] The V-2, however, had two advantages over the WAC Corporal. First, because the V-2 functioned primarily as a weapon of terror – to be known and feared – it had the authority that came with being the established and internationally known pioneer rocket. Second, having been designed to propel one ton of amatol explosive onto civilians in London, von Braun's creation was engineered to do some heavy lifting. The instrument bay vacated by the warhead could not only hold a weighty scientific payload; it would also encompass an instrument of an altogether different order: the WAC Corporal itself. The idea came from Malina's friend and colleague Martin Summerfield, having recently moved from Aerojet back to JPL. Relations between the German and JPL teams were cool but Colonel Holger Toftoy, the original architect of Operation Paperclip, doubtless thought that such a collaboration would be to everyone's benefit. The new Bumper WAC Corporal, so called as the V-2 would give a 'bump' to the WAC Corporal, would by no means be the first staged rocket. (In a sense the WAC Corporal was already a staged vehicle getting a bump from its solid propellant Tiny Tim booster). But the Bumper would be the first high-altitude platform for a staged rocket. It could test vehicle separation at very high velocity as well as get data on rocket performance at speeds and altitudes hitherto unexplored.

Though the WAC Corporal might itself be considered the instrumental payload of a V-2, inside the smaller rocket was another set of instruments, albeit limited to 50 lbs in weight. A Doppler receiver/transmitter could determine the speed and position of the missile while the telemetry system could relay in flight data such as the nose cone temperature – as yet unknown information about the behaviour of the rocket under hypersonic conditions. Of the six Bumper launches at White Sands, only Round 5 was successful but the extent of its success eclipsed the programme's other failures. On 24 February 1949, after it reached 244 miles at Mach 6, the Bumper acquired public attention for becoming the highest and the fastest human object to date.[55] In the absence of a rival space programme, however, this achievement was not thought to herald a new era. Years later, Malina would complain to the Smithsonian's Frank Winter that 'if those who publicise such matters had not been asleep in our country' then 'a reasonable claim could have been made ... that the Bumper WAC project opened up the Space Age well before the Sputnik'.[56] That glory went to the Soviet Union leaving the 'beep' of Sputnik's telemetry to become the sound that haunted America. But the first use of radio equipment at extreme altitudes, indeed the first demonstration that

---

54   William Corliss, *NASA Sounding Rockets, 1958–1968: A Historical Summary* (Washington: NASA, 1971), 13; Stanley O. Starr, 'The Launch of Bumper 8 from the Cape, the End of an Era the Beginning of Another', paper presented to the 52nd Astronautical Congress, http://www.nasa.gov/pdf/171684main_Bumper8.pdf, accessed 1 August 2014.

55   Clyde T. Halliday, 'Seeing the Earth from 80 Miles Up', *National Geographic* 98, (October 1950), 511–28.

56   Letter from Malina to Frank H. Winter, 16 March 1979, Box 6 Folder 34, FJM Collection.

such communication could be sustained through the ionosphere, was in Bumper rather than Sputnik.[57] The success of this radio communication would be extended considerably in the development of the Corporal – the ultimate object of the ORDCIT contract.

Conclusion

In ordinary circumstances, the success of Round 5 of the Bumper WAC Corporal might have occasioned a moment in the limelight for Frank Malina. He would have had to modestly acknowledge the teamwork at JPL, the institution he founded, and pay tribute – albeit through gritted teeth – to Wernher von Braun and the Paperclip group overseeing the V-2 as part of Project Hermes. But in 1949 circumstances were far from ordinary: Malina had already left JPL, rocketry and the United States for a job in Paris at the natural sciences section of the fledgling UNESCO. Exhausted by the demands of rocket development, and dismayed at its growing weaponisation, he had for a long time discussed another career with Liljan. As early as September 1944, Malina told her of his desire 'to back to school for a year or two to study human beings'; 'I would rather be a failure in that sphere of human endeavor than a recognized success at the end of my life in the field I am now working in'.[58] Though he did not go back to school, he did pursue a life within and across the humanities and the sciences. As a pioneer of modernist kinetic painting, and the founding editor of *Leonardo* journal, he need not have worried about underachievement in his chosen field. The recognition of his engineering success in his home country, however, was another matter.

The American space establishment was happy to overlook Wernher von Braun's SS-membership, giving him every conceivable honour, from an eponymous crater on the moon to naming a 10,000-seat civic centre in Huntsville, Alabama. Malina by contrast was honoured by the persistent attention of the FBI and the US state department: raiding his home, following him overseas, denying him his passport and, eventually, getting him sacked from UNESCO.[59] Anti-fascist activities and networks that Malina had held in the late 1930s were, by the late 1940s, the subject of intense security interest. Malina's friend Sidney Weinbaum, who had left Linus Pauling's lab to work in the airline industry only to join the growing workforce at JPL, found himself grilled by the FBI in 1949. In the weeks after the Bumper flight, Sidney's wife Lina wrote to Frank that 'we missed greatly your name which should have been in great print at the last rocket experiments … without you it just would not have taken place'.[60] Months later, Weinbaum was sacked from JPL and

57   Stanley O. Starr, 'The Launch of Bumper 8', 11.

58   Letter to Liljan Malina, 11 September 1944, MFA.

59   Much of this is detailed in Malina's FBI file: http://vault.fbi.gov/frank-malina, accessed 15 May 2014.

60   Letter from Lina Weinbaum to Malina, 21 March 1949, MFA.

spent years in jail for perjury, having failed to declare communist involvement on a security questionnaire. Martin Summerfield, who first thought to conjoin the V-2 and the WAC, was denied his security clearance to work on classified projects. Tsien, another friend of Weinbaum, not only lost his clearance but was eventually deported to his native China where he founded the Chinese space programme.[61]

Though it has not been my primary purpose to detail Malina's comparative exclusion from both the historiography and memorialisation of American spaceflight, there remains something significant about the reluctance to mark his contribution through, say, naming a small feature of Martian or Lunar topography. After all, von Kármán has craters on both the Moon and Mars, as well as the Kármán Line marking the boundary between Earth's atmosphere and outer space. If you count Von Kármán Avenue in Orange County, California, his name has been written on three celestial bodies. Even Jack Parsons has a crater, appropriately enough for a devotee of Aleister Crowley, on the 'dark' side of the Moon. Malina's absence is all the more curious since the institution he founded, JPL, one of the world's leading centres for scientific exploration, is now opening up new horizons of Martian territory.

The fact that Malina has no memorial is but one aspect of a wider historiographical tendency that has shaped the conventional story of how, and indeed why, humans have cast their instruments beyond the boundaries of Earth. We need, then, to be alert to the ways in which lingering Cold War anxieties have obscured the significance of work that created the enabling conditions for the development of the space sciences and for extra-terrestrial exploration. The story of Frank Malina is important in recalling the motivations that took humans out of their planetary home for the first time: a civilian desire for scientific knowledge realised through the outworking of military competition inside the frame of Westphalian territoriality. The philosopher of social science Rene Girard coined the phrase 'mimetic desire' to describe how all desire – in this case, a desire for a vehicle of transcendence and destruction – is a kind of contagion; a desire for an object arises not from itself but in imitation of a third party who holds the same desired object.[62] In this context, it is hard to overlook the fact that the German V-2 was initiated on hearing (as it turned out, false) rumours that the Americans were developing a rocket weapon. In America, Malina's work found support only when intelligence reports of the V-2 found their way to the Army Air Force. Such triangulations abound. That evidence of the human impulse to explore was mediated through both scientific and military objectives is reflected in the constitution of the rocket itself: at once instrument, vehicle, and projectile. Even the idioms of exploration and targeted destruction are entangled: James W. Bragg,

61   Iris Chang, *Thread of the Silkworm* (New York: Basic Books, 1995); Zuoyue Wang, 'Transnational Science during the Cold War: the Case of Chinese/American Scientists', *Isis* 101 (2010), 367–77.

62   Rene Girard, *Deceit, Desire, and the Novel: Self and Other in Literary Structure* (Baltimore: Johns Hopkins University Press, 1976).

the the Army's official chronicler of the Corporal missile, asked that 'it not be forgotten ... that the somewhat rude, rough, uncouth pioneer Corporal blazed the trail through a wilderness of dynamics, aerodynamics, and electronics ... pointing out the path for manufacturers and military personnel to follow with the designing, fabrication, and operation of more refined, sophisticated second and third generations of such missile weapon systems.[63]

Frank Malina is not only important for his theoretical and institutional legacies but also for his involvement in establishing the International Academy of Astronautics (IAA) and, thus, for giving shape to astronautics as a disciplinary field. And here the practice of instrumental measurement is ceded a particular importance. Writing to von Kármán about the draft statutes of the IAA, Malina made the distinction between instrumental procedures necessary for the development of astronautical engineering on the one hand, and the development of astronautics in order to bear the instruments of 'pure' scientific endeavour on the other. 'Individuals in ... the basic sciences of physics, chemistry etc. may use the developments in the field of astronautics such as vehicles which can carry instruments for learning new knowledge of nature, but this does not make the individual who uses the instruments an astronautical scientist'.[64] In practice, however, I wonder if these are not harder to distinguish. Malina and his fellow rocketeers were, I would argue, among the first true explorers of space even if they only sought to enable, rather than personally conduct, meteorological investigations or cosmic ray research. A similar argument might, of course, be made about Galileo – an unacknowledged space explorer bringing back images of the celestial realm through the new technology of the telescope. What distinguishes these engineers is that for the first time, rather than prosthetically extending the terrestrial observer, they could locate their instruments within the field of exploration – beyond the envelope of Planet Earth.

---

63   Bragg, *Development of the Corporal*, x.
64   Letter from Malina to Theodore von Kármán, 11 July 1959, Box 2 Folder 8, FJM Collection.

Chapter 12

UK Radar (Dis)Integration in the 1960s: Linesman/Mediator Radar Development and the Calculus of Nuclear Deterrence

Graham Spinardi

Technology has long played a role in making the world a smaller place, through transportation, communication, or surveillance. What Donald Janelle refers to as 'time-space convergence' and David Harvey as 'time-space compression' was particularly significant in military terms during the early Cold War.[1] Technologies initiated in World War II changed the scale of warfare in several ways. The ballistic missile presented an almost unstoppable threat that could travel thousands of miles in minutes. The atom bomb, and, later, the hydrogen bomb, meant one relatively small package could lay waste to a city. And radar provided an instrument that could enable above-the-horizon threats to be spotted at great distance.

These three technologies dominated UK air and nuclear defence policy after 1945. A fading 'Great Power', the UK sought to maintain its security and status in the face of a Soviet nuclear threat that became increasingly potent during the 1950s and 1960s. In particular, the advent of Soviet thermonuclear weapons led UK planners to doubt whether their small, highly-populated islands could realistically fight a nuclear war.[2] The destructive power of such weapons meant that just one detonation could devastate major cities. These concerns were amplified by the development of ballistic missiles that could reach the UK from Eastern Europe in less than four minutes, and against which no effective defence was thought feasible.[3]

The late 1950s thus saw a major shift in UK planning for nuclear war. No longer was defence considered possible; future strategy was to be based around

---

1   Donald G. Janelle, 'Spatial Reorganization: A Model and A Concept', *Annals of the Association of American Geographers*, 59 (1969): 248–64; David Harvey, *The Condition of Post-Modernity: An Enquiry into the Origins of Social Change* (London: Blackwell, 1989).

2   Melissa Smith, 'Architects of Armageddon: the Home Office Scientific Advisers' Branch and Civil Defence in Britain, 1945–68', *British Journal for the History of Science*, 43 (2010): 149–80; Matthew Grant, 'Home Defence and the Sandys Defence White Paper, 1957', *The Journal of Strategic Studies*, 31 (2008): 925–49.

3   Jeremy Stocker, *Britain and Ballistic Missile Defence, 1942–2002* (London: Frank Cass, 2004), 58.

deterrence. UK nuclear policy followed the 1946 dictum of the American strategist, Bernard Brodie: 'Thus far the chief purpose of our military establishment has been to win wars. From now on its chief purpose must be to avert them'.[4] This approach had implications for UK equipment and planning. Against an unstoppable nuclear missile threat it was thought that there was no need for the large numbers of interceptor aircraft and surface-to-air missiles then being procured. Nor was an extensive radar system required to detect enemy bombers and to direct aircraft and missiles against them.[5] The challenges presented by changing technology were compounded by the rate of change. The problem for the UK was that competing in military technologies was becoming financially untenable, quite apart from the difficulty of finding sufficient scientists and engineers.

These twin consequences of rapid technical change – the shift from defence to deterrence, and the challenge of affordability – found a common focus in decisions that were made about UK air defence at the end of the 1950s. The UK's air defence radar system would no longer be based around the need to prevent a bomber attack, but would instead be geared towards defending the Fylingdales Ballistic Missile Early Warning System (BMEWS) from being rendered ineffective by jamming aircraft – thus, it was thought, maintaining the credibility of the UK deterrent.[6] Without adequate warning, Britain's V-bomber based nuclear weapons would be vulnerable to attack before they could become airborne. The operational security of the BMEWS was thus considered essential to the credibility of the deterrence strategy, as well as being key to the organisational interests of the Air Force. At the same time, it was thought that economies could be achieved by combining air defence and air traffic control (ATC) facilities by merging a new air defence radar system ('Linesman') with a new air traffic control system ('Mediator'). This made sense as regards the shared use of radar hardware. What proved unexpected – and unexpectedly difficult and costly – was the shift from hardware to software as central elements of technological performance. Moreover, the expected commonality over technology did not extend to the interests or behaviour of the organisations involved. By the late 1960s, almost all the key premises of Linesman had proven unfounded. Changes in strategic thinking and the perception of the threat undermined the fundamental assumption that a vulnerable, centralised system was suitable. The benefits of commonality of air defence and air traffic

---

4   Bernard Brodie (ed.) *The Absolute Weapon: Atomic Power and World Order* (Harcourt, Brace and Company, 1946), 76.

5   Detailed accounts of post-war UK radar developments can be found in Jack Gough, *Watching the Skies: A History of Ground Radar for the Air Defence of the United Kingdom by the Royal Air Force from 1946 to 1975* (London: HMSO, 1993); and R.H.G. Martin, *A View of Air Defence Planning in the Control & Reporting System 1949–1964* (Oxford: Alden Press Ltd, 2003).

6   Graham Spinardi, 'Golfballs on the Moor: Building the Fylingdales Ballistic Missile Early Warning System', *Contemporary British History*, 21 (2007): 87–110.

control were not achieved. The ambitious informational focus of Linesman exceeded British capabilities in computation and software engineering.

In this history, radar may have been a powerful instrument for scanning the sky, but its use was purely pragmatic, being shaped by strategic and organisational goals, rather than by the pursuit of knowledge *per se*.[7] What in 1940 was first named Radar (Radio Detection and Ranging) came to practical fruition in World War II.[8] Typically, pulses of radio waves are transmitted from an antenna and the reflections that bounce back can be used to determine the location and velocity of the reflecting object. Radar's origins reflected its military value. Once weapons took to the skies, in the form either of aircraft or, later, missiles, the ability to detect them could provide early warning and perhaps a chance of interception. Detection, however, is not the same as identification. The resulting 'blips' seen on radar screens have no self-evident meaning. Early radars were not only prone to picking up the mass movements of birds (the so-called 'angels' first reported during World War II), but also the interpretation of observed blips required further knowledge.[9] As the skies became more crowded what mattered was not simply that an aircraft had been spotted, but to know the kind of aircraft and its nationality.

The use of radar as an instrument in the Cold War constituted another element of the military's desire to map space. Much of this effort was devoted towards mapping geophysical properties.[10] Radar's primary role was different in that the space to be mapped – the sky – was constantly changing as objects moved

---

7   The development of that related major project aimed at knowledge collection, the Jodrell Bank radio telescope, featured a similar range of pragmatic political interests: see Graham Spinardi, 'Science, Technology, and the Cold War: The Military Uses of the Jodrell Bank Radio Telescope', *Cold War History*, 6 (2006): 279–300.

8   For the origins of radar, see Robert Buderi, *The Invention that Changed the World* (London: Simon & Schuster, 1996).

9   David Clark, 'Radar Angels', *Fortean Times*, 195 (2005). Downloaded 2 September 2014 from http://drdavidclarke.co.uk/secret-files/radar-angels/; see also Helen Macdonald, '"What Makes You a Scientist is the Way You Look at Things": Ornithology and the Observer, 1930–1955', *Studies in History and Philosophy of Biological and Biomedical Sciences* 33 (2002): 53–77.

10   John Cloud, 'Crossing the Olentangy River: The Figure of the Earth and the Military-Inustrial-Academic-Complex, 1947–1972', *Studies in the History and Philosophy of Modern Physics* 31 (2000): 371–404; Deborah Jean Warner, 'From Tallahassee to Timbuktu: Cold War Efforts to Measure Intercontinental Distances', *Historical Studies in the Physical and Biological Sciences* 30 (2000): 393–415; Kai-Henrik Barth, 'The Politics of Seismology: Nuclear Testing, Arms Control, and the Transformation of a Discipline', *Social Studies of Science*, 33 (2003): 743–81. On mapping the ocean floors, see Graham Spinardi, *From Polaris to Trident: The Development of US Fleet Ballistic Missile Technology* (Cambridge: Cambridge University Press, 1994). For gravity mapping, see Donald MacKenzie, *Inventing Accuracy: A Historical Sociology of Nuclear Missile Guidance* (Cambridge, MA: MIT Press, 1990). See also Simone Turchetti and Peder Roberts (eds) *The Surveillance Imperative: Geosciences during the Cold War and Beyond* (New York: Palgrave Macmillan, 2014).

across the area scanned. This space needed to be mapped in near real time, but was dynamic in nature. While the instrumental properties of radar provided traces on screens that operators could match to aircraft, the significance of their ability to observe airborne objects was determined by broader strategic considerations. What constituted a 'threat' was not defined by what appeared on radar screens but, rather, by how prevailing strategy viewed the likely nature of warfare in the nuclear age. When Linesman was first conceptualised, its role depended on distinguishing enemy from friendly aircraft. By the end of the 1960s, the role of air defence radar had returned to its more traditional framing whereby early warning and interception coordination was key. It again became conceivable to fight a limited war that might gradually shift from conventional to nuclear; deterrence was reframed as dependence upon the willingness to do so.

Post-War Air Defence and Air Traffic Control

Despite the key role played by radar in UK air defence during World War II, the aftermath of the war was a period of neglect.[11] It took until 1951 before Sir Henry Tizard, the chairman of the Defence Research Policy Committee, was able to raise the status of Britain's air defences.[12] The programme to modernise Britain's early warning system ('Operation Rotor') was then given 'priority over all other defence work, except atomic energy and guided missiles'.[13] Rotor Stage 1 was handed over to Fighter Command in April 1956. This involved the installation of radars at 39 new stations, providing full coverage of the east and south-east coasts – full coverage, that is, in business hours only as the stations were unmanned overnight and at weekends.[14] However, no sooner was Rotor completed than it was considered obsolete. The wartime approach, continued in Rotor, was to have two types of radar. Early warning radars were known as CH after the wartime 'Chain Home' that had been built along Britain's southern and eastern coasts to warn of German attack. Another radar system – GCI for 'Ground-Controlled Interception' – was used to coordinate air defence aircraft based on the early warning data.

One of the significant characteristics of the new Type 80 radars, however, was that the Type 80 Mk 3, intended for the GCI role, had longer range than the Mk 1, which was replacing the CH radars in the early warning role. To provide the earliest possible warning, the CH radars had been installed along the coastline. The GCI radars were located inland, taking advantage of suitable saucer-shaped

---

11   Gough, *Watching the Skies*, 52.

12   Stephen Twigge and Len Scott, *Planning Armageddon: Britain, the United States and the Command of Western Nuclear Forces 1945–1964* (London: Harwood, 2000), 270.

13   Memorandum to Minister of Defence, 10 December 1951. The National Archives PREM 11/72. [Hereafter, all references to AIR and DEFE are to files in The National Archives.]

14   Gough, *Watching the Skies*, 153–4.

topographical depressions to enhance their performance. Because the coastal early warning Type 80 Mk1 turned out to have no range advantage over the inland GCI Mk3, it seemed logical to do away with the separation between GCI and CH stations, and to use one radar system for both roles.[15] It was this potential for combining the early warning and fighter control functions into single radar stations that formed the basis for the '1958 Plan' that was developed during 1956 and 1957. This plan divided the UK into nine geographical sub-sectors, each with a 'comprehensive' radar station based around the Type 80. Each sub-sector would have been essentially autonomous. The combination of the early warning and GCI roles was intended to eliminate those information flows that had previously caused delay and error.[16] Air defence envisaged an increasing number of Surface to Air Guided Weapons (SAGW) operated in conjunction with fighter aircraft and providing protection of the UK nuclear deterrent (V-bombers were first deployed in 1955 and a land-based Blue Streak missile was planned) as well as some city protection.[17]

Even as it was being formulated, this approach was being undermined by the threat posed by a jamming technology known as the carcinotron. Originally invented in 1950, this microwave oscillator used a vacuum tube to produce radio waves. Unlike the magnetron that had transformed the portability of radar in World War II, the carcinotron was tunable over a wide frequency band. Trials carried out in late 1954 appeared devastating for the future of radars such as the Type 80, since even a carcinotron of modest power could hide the radar profile of an aeroplane carrying it.[18] Initially, the only solution appeared to be a radar with a very high power transmitter and a very large aerial. The Royal Radar Establishment (RRE) worked on a design code-named 'Blue Riband', drawing on the experience of the Manchester University Radio Telescope at Jodrell Bank with its 250 foot diameter aerial.[19]

A more elegant solution, however, was devised in the form of the evocatively-named Winkle passive detection system: this relied on correlation of data from two or more radar locations to overcome jamming.[20] The independent radar stations proposed in the 1958 Plan thus no longer made sense. Integration of data between stations was also desirable in order to avoid the expensive replication of equipment, and because the overhead cover provided by the new Type 85 radar was inferior to the Type 80 (and its L-band variant, the Type 84) and so needed the help of neighbouring radars.[21] The idea of autonomous comprehensive radar stations was thus abandoned in favour an integrated system with data sent

---

15 Ibid., 150–51.
16 Ibid., 154.
17 Martin, *View of Air Defence Planning*, 36–41.
18 Gough, *Watching the Skies*, 157–8.
19 Ibid., 170–72.
20 Ibid., 185.
21 Martin, *View of Air Defence Planning*, 45.

to central control centres. This approach, known as 'Plan Ahead', was approved by the Air Council in January 1959.[22] Plan Ahead involved three stations along England's east coast, each equipped with a Type 84 search radar, and a Type 85 radar giving bearing, distance and height. Data from these three stations, along with the passive detection system, would then 'be transmitted to a Master Control Centre in processed form suitable for the new techniques of automation which are being applied to air defence'.[23] The costs of completing Plan Ahead were estimated to be between £54M and £73M.[24]

A fundamental issue in the conception of Plan Ahead was the significant shift in defence policy brought about by the 1957 Sandys Defence White Paper. This emphasised the missile threat, with the main purpose of air defence to 'convince the Russians that they cannot attack the United Kingdom's nuclear retaliatory forces with any assurance of neutralizing them and thereby avoiding nuclear retaliation', as well as 'to minimize the effect of enemy attack', prevent enemy aircraft intrusion, and reassure allies.[25] It was soon realised, however, that in a time of financial stringency the defensive strategy assumed for Plan Ahead could not be justified in the face of thermonuclear-armed ballistic missiles. In 1960 the idea of air defence, as traditionally conceived, was scrapped. It was decided 'to abandon the defence of the deterrent and to limit air defence in the United Kingdom to the prevention of intrusion and jamming'. The requirements now were to '(a) provide early warning of hostile aircraft equipped with modern jamming devices; (b) give an accurate picture of the activities of enemy aircraft in order to help the Government to decide upon the appropriate political or military reaction; and (c) enable our fighters to deter and, if necessary, intercept intruding or jamming aircraft, of which some might be supersonic and fly at very great heights'.[26] Military air defence became an air traffic control challenge.

This convergence between military air defence requirements and civil air traffic control came at a time when the civil system desperately needed modernisation. Greater numbers of civil aircraft, and the increasing use of higher altitudes by the new 'jet' airliners, meant that there were more potential conflicts with military aircraft.[27] Air traffic control was facilitated by 'an airways system within which all aircraft are separated from each other in time or space and are protected from

22   Gough, *Watching the Skies*, 186.

23   'Cabinet–Integration of Air Defence and Air Traffic Control Radar Plans', Memorandum by the Minister of Aviation, Second Revised Draft, October 1961. AIR 2/16027.

24   Ibid.

25   Martin, *View of Air Defence Planning*, 59.

26   Draft paper for Cabinet Defence Committee, 'Integration of Air Defence and Air Traffic Control Radar Plans for the United Kingdom', attached to letter from R.C Kent, AUS(A) to DUSI, 20th August, 1962. AIR 2/16367.

27   'Linesman/Mediator: Note on the Current Air Traffic Control System and the Problems Associated with is Development', attached to letter from M.T. Flett, DUSI to PS to S of S, 19/10/62. AIR 2/16367.

other external traffic because the latter is not allowed to fly into an airway except under specially arranged procedures designed to ensure safety'.[28] In the early 1960s most aircraft in UK airspace flew outside of these controlled airways, but the situation was changing rapidly as commercial aviation increased and military operations declined.[29] As Vice Chief of the Air Staff for Signals, G.C. Eveleigh put it, 'Jet airliners were filling the airspace previously regarded as the military's playground'.[30] The efficient operation of jet airliners required them to fly at high altitude, so the airways were extended upwards into what previously had been the preserve of the military. To help counter the danger of collisions due to aircraft 'penetrating the airways system, the Air Ministry progressively brought into service a number or redundant air defence radar stations'.[31] These Air Traffic Control Radar Units (ATCRU) improved air safety. Studies undertaken in 1961 showed that 'air miss incidents' were between 6 and 14 times higher in airspace lacking radar cover.[32] Despite some uncertainty over the data it was concluded that the ATCRU's 'limited service has already been justified by the use made of it and by the significantly lower rate of airmisses recorded among controlled traffic than among other traffic'.[33] However, the ATCRUs only provided coverage in some areas, and were considered 'obsolescent and, more important, cannot be relied upon in bad weather because they "see" the clouds'.[34]

With ATC radar cover lacking or limited in many areas, increasing potential for military/civil air traffic conflict, and reliance on manual procedures for tracking aircraft, it was clear that modernisation was necessary. The matter resulted in a number of reports by the late 1950s and the establishment of the Air Traffic Control Board in 1959, and, in turn, to the National Air Traffic Control Planning Group (known as the 'Patch Committee' after its chairman, Air Marshal Sir Hubert Patch).[35] With both civil and military radar having the same basic objective – detecting, locating, and identifying aircraft – the question now arose of whether the same instruments could be used to provide a common solution.

---

28   Ibid.

29   Ibid.

30   Foreword, Martin, *View of Air Defence Planning*.

31   'Linesman/Mediator: Note on the Current Air Traffic Control System and the Problems Associated with is Development', attached to letter from M.T. Flett, DUSI to PS to S of S, 19/10/62. AIR 2/16367.

32   'Reduction of Air misses in Areas covered by ATCRU's', D.P.J Smith, ATC3 (Plans) to F1 (Mr Day), 5th April, 1962. AIR 2/16367.

33   B. M Day to DUSI, 5th April, 1962. AIR 2/16367.

34   'Linesman/Mediator: Note on the Current Air Traffic Control System and the Problems Associated with is Development', attached to letter from M.T. Flett, DUSI to PS to S of S, 19/10/62. AIR 2/16367.

35   Gough, *Watching the Skies*, 220–22.

From Plan Ahead to Linesman/Mediator

Britain's Air Ministry approved Plan Ahead in May 1960, but, sceptical that the future air threat justified such expenditure, Prime Minister Harold Macmillan only agreed to Plan Ahead on the condition that other air defence projects, such as the balloon-borne Blue Joker, were cancelled.[36] Discussion continued about the relevance of Plan Ahead to defence requirements, and whether ways could be found to reduce its cost. Solly Zuckerman, Chief Scientific Advisor to the Ministry of Defence, concluded that 'It is clear to me that the only course which offers any prospect of appreciable savings and, indeed, the right course on merits, is to design a single system to cover the needs of both defence and civil aviation'.[37] On 5 December 1960, the Minister of Defence presented a paper to the Cabinet Defence Committee advocating a joint system for air defence and air traffic control. This was approved and, on 7 December, the Cabinet Defence Committee 'invited the Minister of Aviation, in consultation with the Secretary of State for Air, as a matter of urgency, to initiate a study of the possibilities of a joint radar system for air defence and for civil air traffic control'.[38]

On 16 December 1960 presentations were given to the Patch Committee at the Royal Radar Establishment (RRE) describing the scenarios facing Britain's civil air traffic control, military air traffic control, and air defence. The conclusion was that a joint system could be readily accomplished, the only significant change being the requirement to co-locate the Southern Air Traffic Control Centre (SATCC) and Plan Ahead's Master Control Centre. The MCC would therefore need to be located in the London area, rather than at Bawburgh in Norfolk as was originally planned.[39] An Assessment Group was set up comprising representatives of the Ministry of Aviation and the Air Ministry, chaired by Sir Laurence Sinclair, Controller of the National Air Traffic Control Service (CNATS). Its conclusion was 'that the present air defence and air traffic control plans can be combined to form an integrated system of radar coverage over the United Kingdom'.[40]

This combined approach seemed logical. Not only was there little appetite to pay for a dedicated system that now only had a limited air defence role, but also the remaining role depended upon identifying intruder aircraft (potential BMEWS jammers) and so was essentially an air traffic control rather than early warning challenge. What mattered now was not just that blips could be spotted on radar screens, but also that those blips could quickly be identified as friendly. They would otherwise be presumed to be intruders, and the efficacy of the system

36   Ibid., 187–8.
37   Quoted in ibid., 189.
38   'Linesman/Mediator', F. Wood, DUS(1) to PUS, 20/10/61. AIR 2/16027.
39   Gough, *Watching the Skies*, 220–22.
40   'Cabinet–Integration of Air Defence and Air Traffic Control Radar Plans', Memorandum by the Minister of Aviation, Second Revised Draft, October 1961. AIR 2/16027.

depended on the ability to process data on known aircraft, identified by knowledge of their flight plans and data from secondary radar beacons, as much as it did on radar coverage. As noted within the Air Ministry, without such data integration, 'radar in itself merely provides a crude picture of events outside; a target may be seen as a luminous mark on a screen, but there is nothing there to tell what it is, where it is going, or who is responsible for it'.[41]

The combined ATC and air defence plan was accepted by the Air Ministry in January 1961, and new codenames were assigned; Linesman for air defence and Mediator for ATC.[42] The Treasury 'approved the revised scheme in principle' allowing equipment production and works services to proceed.[43] The Linesman component involved the three main radar stations in conjunction with a passive anti-jamming system feeding data to the MCC at West Drayton. Yet, Linesman/ Mediator faced many challenges. Despite the attempt to reduce costs by combining the military and civil radar requirements, costs remained high. Apportioning the share of costs between two government departments would prove a challenge. On the military side there remained serious reservations by many who doubted the approach taken, a debate complicated by the difficulties of adapting to the changing, and increasingly limited role, of air defence in an age of nuclear missiles. While strategists considered that fighting a war against the Soviet Union was inconceivable (because of the expectation that nuclear devastation would rapidly ensue), the Royal Air Force, and particularly Fighter Command found such a reduction of its role unpalatable.

Integration or Autonomy? The Limits of Air Defence

Initial objections to Linesman from the Royal Air Force centred on the plan to change the location of the MCC from Bawburgh in Norfolk to West Drayton near Heathrow. The concern (prescient, as it turned out) was that the London MCC would not be underground, and would, therefore, be vulnerable to attack. Given that current policy envisaged a three-day conventional war (which would then either end or become a nuclear conflict), and that the radars themselves were above ground and also considered vulnerable, the vulnerability of the MCC was not itself considered a weak link. Steps could also be taken to reduce reliance on the MCC by providing the radar tracking stations with the capacity to extract data locally and to have control of fighter aircraft.[44] This issue of local autonomy versus centralised data integration caused much concern for Fighter Command, who feared that development of a large centralised system would lead to serious

---

41   Draft 'Notes on the Linesman/Mediator System' attached to Loose Minute from B.M. Day to DUSI, 11th April, 1962. AIR 2/16367.

42   Gough, *Watching the Skies*, 224.

43   'Linesman/Mediator', F. Wood, DUS(1) to PUS, 20/10/61. AIR 2/16027.

44   Gough, *Watching the Skies*, 224.

delays in the ability to intercept supersonic intruders. It was argued that the centralised processing capability of the MCC was not only unnecessary and expensive, but also liable to delay implementation of any effective air defence system. As Commander-in-Chief of Fighter Command, H.D. McGregor noted in January 1961:

> To provide early warning information of a manned attack of aircraft equipped with modern jamming devices, it is essential that radars Type 85 and a passive detection system are provided. To meet this specific requirement, however, it is not essential that information derived from the radars and the passive detection system should be correlated at any control centre. The essential factor is speed and the provision of data direct from a raw radar picture ... A delay in the provision of an MCC will therefore not prejudice the provision of early warning information with which to alert our deterrent forces nor will the provision of a control centre improve the quality of the alerting information. However, if plans for the radar stations are delayed, pending the finalisation of the overall plan, the provision of adequate early warning will be delayed to an unacceptable extent.[45]

On 6 December 1961 McGregor wrote to the Vice Chief of the Air Staff (VCAS) arguing against the MCC, and stressing the cost savings to be made from relying instead on three autonomous radar stations. McGregor was concerned that the reduction in the number of master radar stations to three from the seven envisaged in the 1958 plan, and MCCs from two to one had 'made the whole system so brittle and vulnerable from a military point of view that I am convinced that on this count alone it is not militarily justified in the form now proposed unless the brittleness is reduced by the introduction of a moderate degree of control capability at the Main Radar Stations'.[46]

McGregor referred to a February 1961 study 'comparing the relative merits of the centralised MCC concept with the decentralised system of three independent or "autonomous" radar stations' which had concluded that MSR autonomy was needed 'to close the gap' before the expected completion of the MCC. He noted that 'a single MCC relocated remotely above ground, and with lengthy broad band radio links is technically and militarily vulnerable', that it would 'inevitably be "off the air" for one reason or another on frequent occasions and for long periods', thus pointing to 'the urgent need for an "autonomous" system as a standby'.[47] Although conceding that a centralised system 'was desirable to obtain the most effective and economic allocation of air defence weapons and targets in the event of a mass attack on this country', McGregor argued that 'the Directive of the

---

45   'Plan "Ahead"', paper attached to Loose Minute from E.J. Morris, D. of Ops (AD), 24 January, 1961. AIR 2/15782.

46   H.D. McGregor, Air Officer Commanding-in-Chief, Fighter Command to Under Secretary of State, Air Ministry (VCAS), 6 December, 1961. AIR 2/16027.

47   Ibid.

Defence Committee that financial expenditure cannot be justified on equipment which is of use simply for fighting a war, now means that many of the arguments in favour of the centralised system are no longer valid'.[48]

McGregor's reading of the changing role of air defence was idiosyncratic. The shift in policy towards protecting the BMEWS function meant that identifying unfriendly intruders, and not early warning and interception of mass attack, was the key objective of Linesman. To achieve this, air traffic control functions required centralisation because, it was argued, 'military Air Traffic Control requirements are more demanding than those for civil aviation'. This Air Staff official noted that in order 'to satisfy the military Air Traffic Control requirement, something very closely akin to LINESMAN would be required in any case and the only result of putting LINESMAN back in the melting pot now would certainly be to delay – and possibly to jeopardise altogether – the introduction of a scheme which would adequately meet our military requirements'.[49] Nevertheless, the Chief of the Air Staff (CAS) was sufficiently swayed by McGregor to argue that 'we could not conscientiously go ahead with the LINESMAN plan as at present conceived when it seemed certainly to be over-elaborate and costly for our Air Defence needs and it was open to doubt whether it was not similarly so on the Air Traffic Control side, both civil and military'.[50]

The Air Ministry's Director General of Signals (DGS) did not agree. Failure to integrate air defence with air traffic control and to build an MCC would mean 'acceptance of a continuation of the very low quality recognition system now in use and this means that intruders who adopt an intelligent approach will usually escape detection'.[51] This view prevailed. Linesman remained unchanged in accordance with the policy agreed in 1960 for Plan Ahead, which 'was never in fact presented ... on the basis that it was needed to enable the fighter and SAGW forces to defend the deterrent'.[52] The policy was reaffirmed that 'the defence of the deterrent should be abandoned' and that 'the objectives of Air Defence in the United Kingdom ought to be limited to the prevention of intrusion and jamming'.[53] Although the radars themselves looked upwards, exploring the sky for objects, the technical aspects of their design depended on earthbound deliberations over nuclear strategy and machinations fuelled by organisational interests.

---

48   Ibid.
49   R.F. Butler, PS to CAS to DGS, 12 December 1961. AIR 20/10918.
50   Ibid.
51   C.M Stewart, DGS to VCAS, 14 December 1961. AIR 20/10918.
52   R.C. Kent, AUS(A) to A.D. Peck, Treasury, 11 July 1962. AIR 2/16367.
53   Ibid.

Military-Civil Integration and the Issue of Cost

Central to Linesman/Mediator was the idea that sharing radar facilities would save money. It was important, therefore, that the expected savings were realised, with the costs shared between the military and civil organisations. From the onset, however, the joint approach was bedevilled by wrangling between the Air Ministry and Ministry of Aviation (MoA), with each side reluctant to pay for elements of the system that it considered more important to the other. One focus of this dispute in 1961 was the two Type 84 radar stations planned for Winkleigh in Devon and Bishops Court in County Down, Northern Ireland. Some Air Ministry official contested the MoA's view that 'that the installation of the two T.84's at these two locations would be primarily for air defence purposes', arguing that 'the air defence benefits would be quite marginal, and ... we would not have felt justified in leaving these radars on order to meet a purely air defence need'.[54] Others in the Air Ministry were more moderate: 'I would like it made quite clear to MOA that we are not trying to palm the whole bill on to them. We must retain a strong interest to ensure that the right radars are put into Winkleigh and Bishops Court to complete the cover needed for air traffic control in the upper airspace and I am fearful that the civil aviation interests left to themselves will adopt a solution founded on old concepts rather than on the new resulting from the Mediator-Linesman partnership'.[55] A further bone of contention concerned the data handling centres. The Air Ministry bridled at the prospect that defence funding would be used to pay for air traffic control developments. It was noted that 'what stood out from this was the determination of the Air Traffic Control Financiers to attribute as much of the cost of this to Linesman as possible with the expectation that Air Votes would bear the full cost of it and to emphasise the military aspects of the cost of Mediator with a view to getting a substantial contribution from Air Votes towards the cost of it'.[56]

This concern was heightened in October 1961 by steep increases in the estimated cost of providing data handling for both the MCC and the Northern ATCC from £10M to £17M. This led to questions over whether 'all of this data handling equipment is necessary for air defence purposes solely'.[57] Officials were concerned as to whether the MOA was 'working to the proper concept of air defence', and other increases in the cost of military ATC were seen as 'an indication of the manner in which MOA are trying to attribute costs in this paper to military requirements as much as they possibly can'.[58] The Air Ministry's Director of Radio agreed 'that the MOA financiers are determined to palm off as much

54   F. Wood to J.M. Wilson, draft attached to letter from W.E. Dowling, Head of F6 to DGS, 25 September 1961. AIR 2/16027.

55   C.M. Stewart, DGS to Head of F6, 2 October 1961. AIR 2/16027.

56   W.E. Dowling, Head of F6 to D of Radio, 11 October 1961. AIR 2/16027.

57   Ibid.

58   Ibid.

of the cost of providing the civil air traffic control facilities as they can upon the Air Votes'.[59]

Allocating costs was difficult because the central concept of Linesman/ Mediator required integration of data handling systems. The L1 building and the collocated SATCC would both draw on the same data extraction and computing centres. Data handling and flows were complicated also by links with the Northern ATCC. In general 'the equipment is inextricably mixed up and it is not possible … to label parts of it as being there purely for air defence'. Without air defence/ ATC integration, 'a great deal of what is now planned would have to be retained for air defence and doubts would be raised as to whether the expense could be sustained under the current air defence concepts, but this was why Ahead turned in Linesman'.[60] While the disputes within the Air Ministry about autonomy, and between the Air Ministry and the Ministry of Aviation about the division of costs, were going on, the overall cost kept increasing. A joint Ministry of Defence and Ministry of Aviation paper being prepared to gain approval from the Defence Committee took until October 1962 to prepare, with part of the delay due to 'a good deal of difficulty in establishing realistic costings'.[61] The cost of Linesman was now put at £81M. It was noted that 'In fairness this ought perhaps more properly to be compared with the top end of the former bracket (namely £73M) but it is going to be extraordinarily difficult, if not impossible, to avoid the impression that the increase has, in fact, been of the order of £15M to £20M'.[62] The time scale had also changed: 'the date for completion is now put at 1967 which is bound to raise further doubts about the military justification for the scheme'. Furthermore, the £81M did not include autonomy (as yet undecided) for the radar stations, nor the rapidly escalating MOA estimates for civil ATC. Although it was possible to argue 'that the increases would have been inevitable anyway and that the cost of the complete ATC scheme is less than it would have been if it had not been integrated with Linesman, this is not going to make the whole dish more palatable to the Prime Minister and the Chancellor'.[63]

Estimates of increased costs led to considerable debate in late 1961 and in 1962. The Vice Chief of the Air Staff wrote to the Secretary of State in December 1961: 'You will remember that in the latter part of last year we had a considerable struggle over the future of Plan Ahead. Briefly the Prime Minister and the Chancellor took a great deal of convincing that it was right to spend something of the order of £60m (the range of cost was at that time put at between £54m and £73m but for convenience the figure of £60m gained some currency) on an air defence system directed against what would by the time it was available in 1966

59   Draft minute 'Linesman/Mediator', D of Radio to Head of F6, October 1961. AIR 2/16027.

60   Ibid.

61   'Linesman/Mediator', F. Wood, DUS(1) to PUS, 20/10/61. AIR 2/16027.

62   Ibid.

63   Ibid.

be a "minor threat"'.[64] By February 1962 the figure was £125M, leading the Chief of the Air Staff to again press for a simpler system, arguing that 'there seemed to be no reason why a cheaper system could not be evolved giving us the military early warning that we required and also a limited fighter control capability'. He suggested that a 'system based on the proposed new radars together with the northern and southern centres but without automation might be provided within a ceiling of approximately £60M'.[65]

The Air Ministry's Director General of Signals (DGS) countered this claim. His estimates suggested that a minimal system could be had for £55M, but this would only use Type 84 radars and so would be vulnerable to jamming, and would be 'without any data handling other than very elementary track marking'.[66] Moreover, this 'system only includes radars for air defence that fall far below the standard hitherto considered necessary, and a system for military air traffic control which has been repeatedly declared inadequate'.[67] DGS argued that for 'military ATC, with particular reference to the upper airspace, it is essential to have means to co-ordinate the activities of the various controllers' so that the identity of aircraft could be quickly ascertained. His view was that there was 'no way of doing this other than providing a tracking system and data storage that can hold the necessary information against every aircraft in the system and be readily available to every controller. ... Anything less would have neither the tracking nor storage capacity to be useful'.[68] A suggestion made by Sir Laurence Sinclair to save money by delaying data for about three years was also rejected: 'a break in development would mean a delay considerably greater than three years, or else keeping the present development team in being for an extra three years which would cost about the same amount of money'.[69] The other financial alternative – to scrap the passive anti-jamming system – was rejected because a system without it would 'remove our ability to track individual targets and this would mean that we would be unable to assess the strength of any raid and be unable to intercept any supersonic targets'. The passive system was said to add about 30 per cent to the overall cost, but 'improves the ECM capability of the system by a factor of more than 10 and we conclude that it would be unsound, both economically and operationally, to go to the expense of the Type 85s, without backing them with passive detection'.[70] DGS thus argued that significant cost cuttings were impossible: 'Without Type 85s supported by passive detection it would be impossible to analyse with any accuracy the nature of enemy activity if more than three or four jamming aircraft are involved and early warning would thus be unreliable. Similarly, it would not

---

64   VCAS to Secretary of State, 1 December 1961. AIR 20/0918.
65   F.B Sowrey, PSO to CAS to PS to VCAS, 9 February 1962. AIR 20/10918.
66   C.M. Stewart, DGS to VCAS, 1 March 1962. AIR 20/10918.
67   Ibid.
68   Ibid.
69   Ibid.
70   Ibid.

be possible to track individual targets and this would make them at best difficult to intercept, and if the targets are supersonic, impossible'.[71]

While DGS 'repeatedly pleaded that costs should be borne between Air Ministry and MOA on some fair but arbitrary ratio', the problem was that the line between what counted as air defence and what counted as air traffic control was blurred and shifting as the scope of potential integration became better understood and as the significance of aircraft identification for deterrence was recognised. In April 1962 it was noted that 'we would be well advised ... to lay more stress on the service which the plan can give to air traffic control, both military and civil, and less on the purely air defence aspect ... because an examination of the figures suggests that not more than 40% of the total cost of the Plan ... can be ascribed to purely defence needs'.[72]

Separation of Linesman and Mediator

By 1964 it was clear that Linesman/Mediator was not making the expected progress. In March 1964 it was conceded that 'there is now no chance of Linesman at West Drayton being completed either to the contractor's target date of April 1968 or to the Blue Book date of December 1968'. Although the Defence Committee had approved the Linesman/Mediator Plan on 24 October 1962, and 'Linesman requirements in principle were quite clear to RRE then ... 17 months have gone by' and 'Mediator requirements are still not agreed or approved'.[73] Linesman was dependent on progress in Mediator, but Linesman/Mediator meetings early in 1964 concluded 'that it would not be possible to introduce Mediator Stage 2 until 1969 at the earliest', meaning that 'the full Linesman facilities cannot be provided on the planned 1968 "O" date because Linesman depends upon Mediator Stage 2 for such facilities as fighter surveillance and fighter recovery'.[74] Although it had appeared logical to develop a joint military-civil radar system, the practical reality was that while the Linesman plan was well-formed, the technical details for a centralised, automated digital civil air traffic system had only just begun to be considered.

Even the decisions that were made on Mediator lacked substance. Noting the agreement of a new Mediator plan on 29 May, an Air Ministry source complained that 'although it was accepted that the Stage II Plan was issued, in fact it hardly warranted the name of "plan" ... the Plan is so broad that they cannot even begin

---

71   Ibid.

72   Draft letter to J.M Wilson, Ministry of Aviation, attached to note from B.O. Bubbear, PS/DUSI, 6/4/62. AIR 2/16367.

73   'Linesman – Delays', Loose Minute from A.D.S. Phillips, Radio 1, 17 March, 1964. AIR 2/16763.

74   M.K.D. Porter, D.G. Sigs (Air) to VCAS, 29 May 1964. AIR 2/16763.

to translate the Plan into technical detail'.[75] This was corroborated by the Royal Radar Establishment who had 'made an investigation and have agreed that the Mediator Stage II Plan is such a skeleton and so much detail is omitted that they have not been able to discover whether it is compatible or not'.[76] A further problem was that there were 'insufficient qualified staff at RRE to cope with the volume of work for Linesman and Mediator', key personnel having moved to the new European air traffic control agency, Eurocontrol.[77] Many aspects of the system thus could not be specified to allow work to proceed.[78]

It was clear that the whole idea of Linesman/Mediator was fundamentally flawed because the implementation schedules were unsynchronised. There had not been the time or resources available for all the new ATC ideas to be developed and planned. While Linesman was well formulated (based on Plan Ahead), Mediator's 'air traffic requirements were not defined and system planning was practically nil'.[79] This meant that 'the design of Linesman/Mediator, and hence of Linesman, was held up through lack of detailed and approved statements of Mediator requirements'.[80] In May 1964 it was reported that the 'attainment of Linesman/ Mediator as a fully integrated Air Defence and Air Traffic Control System no longer being a possibility within the time scale previously envisaged, I wish to submit for your consideration an alternative implementation plan for Linesman with the object of meeting the air defence requirements by the target date of mid-1968'.[81] Linesman would no longer wait for Mediator, and an interim Stage 1 Linesman capability was planned, with various interim Mediator capabilities to be developed in parallel.

Computing and Software Problems

Even freed from the need to wait for Mediator, Linesman progressed only slowly. The construction of radar hardware went relatively smoothly, but difficulties plagued the development of the software at the heart of the system's Radar Data Processing System (RDPS). The first Type 85 radar at Neatishead, although subject to delays, was handed over to the RAF on 1 June 1967, with the others following at Staxton Wold in January 1968, and Boulmer in May 1968.[82] The same year also saw completion of the passive detection system radars, three of which

---

    75   'Linesman/Mediator–Notes for Informal Meeting, 21 July 1964', A.S. Phillips, Sigs41(Air), 20 July 1964. AIR 2/16763.

    76   Ibid.

    77   'Linesman–Delays', Minute from A.D.S. Phillips, Radio 1, 17 March, 1964. AIR 2/16763.

    78   W. Ross, Chairman, TEECC to CNATCS, 27 April, 1964. AIR 2/16763.

    79   Draft 'Linesman/Mediator'. AIR 2/18311.

    80   M.K.D. Porter, D.G. Sigs (Air) to VCAS, 29 May 1964. AIR 2/16763.

    81   'Script for Wg. Cdr. W.D. Smith Presentation, 19 May 1964'. AIR 2/16763.

    82   Gough, *Watching the Skies*, 262.

were at the Type 85 sites, along with a fourth at Dundonald.[83] The final element of the Linesman set-up was the Master Control Centre housed in the L1 building at West Drayton. By April 1964 it was acknowledged that there was a 'need for an organisation capable of handling the monumental task of programming the computers'.[84] This programming proved a major technical hurdle for the project, and programmers were too few in number. As the project progressed, however, the numbers of computers, and the demands of programming, increased. By mid-1968, all the radar components of Linesman were operational, but the radar data handling system was still delayed by computer problems: it was 'software, not hardware, which comprised the eventual constraint on the attainment of Linesman'.[85] The task was greater than had been foreseen. Extra programmers, including staff from the Atomic Energy Research Establishment, were employed from November 1969, but, by then, the Ministry of Technology had already advised the Air Force Department that Linesman would not be ready until 1973 at the earliest.[86] An additional problem was that the Plessey 090 computers specified at the start of Linesman were significantly underpowered. By January 1969, 'the present capacity of the computer system was less than half of that which would ultimately be required'.[87] More computers were assigned but it was already clear that 'buying in 1973 computers of 1961 design to make use of the redundant peripheral equipment would, we are certain, be completely uneconomic, impracticable and indefensible'.[88]

Changed Nuclear Strategy and the End of Linesman

As computing and software problems mounted, serious consideration was given to the discontinuation of Linesman. The delay in getting Linesman operational was brought into public focus by *The Observer* on 13 October 1968.[89] Having taken years to build, the system was in danger of being obsolete before being deployed, and looked likely to have limited functionality. Yet admitting this, and writing off the investment, was difficult. As the Minister of Defence for Equipment noted in December 1968: 'The alternative to rounding off LINESMAN is to admit publicly

---

83   Ibid., 263.

84   C.M. Stewart to Dr W.H. Penley, DCL, Ministry of Aviation, 8 April 1964. AIR 2/16763.

85   Linesman Progess Committee (L1 Subsystem aspects), Minutes of 46th Meeting 29 August 1968. AIR 2/18194.

86   Gough, *Watching the Skies*, 284.

87   'Linesman Mediator Steering Committee Minutes of 39th Meeting, 9 December 1968', 3 January 1969. AIR 2/18484.

88   'Proposed rewrite of para 5 of DA/ATCS draft letter', attached to letter from AD/LM(T) to Head of F6 (Air), 30 December 1968. AIR 2/18494.

89   Rex Malik, 'Britain's New Early Warning System Already Out of Date', *The Observer*, 13 October 1968. For a more favourable report, see Anthony Shrimsley, 'Bomber-Proof Shield at Britain's Front Door', *Sunday Mirror*, 19 January 1969.

that we have wasted £80M and that we consider it to be unnecessary to provide any significant air defence of the UK against manned aircraft. I believe that such an admission, even if it could be justified on other grounds, would cast such doubt on our defence effort both at home and in NATO that it should not be contemplated. I am, therefore, satisfied that the LINESMAN project should go on'.[90]

A review undertaken by an Air Force Signals Group working party, headed by Air Vice-Marshall Moulton, concluded that there was 'no prospect that the central Linesman complex at West Drayton could be completed in the form hitherto specified by the end of 1973; indeed they doubted if it could be completed at all, in this form, unless more modern machines were substituted for the present RDPS computers'.[91] Ironically given earlier discussions about autonomy, it was now recommended that the central computing problem could be sidestepped by '[d]elegating to the radar stations responsibilities for actual control of interceptions' since 'there is clearly no argument for seeking to complete the project in its original form which would require very considerable extra cost and a very much greater delay in time'.[92] In October 1971 the Vice Chief of the Air Staff requested more studies to 'determine the effectiveness of the UK air defences for three alternative assumptions about the future of Linesman. a. That we spend no further money, but use the system as it. b. That we not only spend no further money, but also make no use of the L1 Complex. c. That we proceed to complete the Linesman system and make full use of it'. [93]

The Vice Chief decided to support the completion of Linesman. Ironically, given the software engineering difficulties that had been experienced, one reason given was that completion was 'a matter of considerable importance to the status and morale of the British electronic industry, particularly that concerning computer software'.[94] The final recommendation was that Linesman should be completed, but further air defence requirements should be considered. This was approved by the Minister of Defence in December 1971.[95] Although the basic Linesman set-up was to be retained, extra radars were to be added to fill gaps in coverage in the north-west and south west of the UK. Even so, it was acknowledged privately that even if Linesman had functioned as specified, the whole concept was fundamentally flawed given the changing strategic requirements of the late 1960s. Linesman's primary role was to spot potential BMEWS jamming intruders given that traditional air defence had been abandoned in the face of the threat of ballistic missiles armed with thermonuclear warheads. Deterrence was considered

---

90   Minister of Defence for Equipment to Secretary of State, 12 December 1968. DEFE 26/226.

91   'Linesman', VCAS to Minister, 16 February 1970. AIR 2/18484.

92   Ibid.

93   VCAS to ACAS(Pol) and ACAS(Ops), 8 October 1971. AIR 20/12429.

94   'Air Force Board, The Future of the Linesman System', Note by VCAS, 1 December 1971. AIR 20/12429.

95   Gough, *Watching the Skies*, 300.

to depend on the ability of the Fylingdales BMEWS early warning radars to alert the V-bombers so that they could take off before being destroyed. This at least was the official rationale of Linesman in the 'closed world' of Cold War nuclear logic.[96] By 1967, when NATO adopted the new nuclear strategy of 'flexible response', this closed world – if not the real world – had irrevocably changed. The logic of flexible response was that the 'defence seeks to defeat the aggression on the level at which the enemy chooses to fight' or 'seeks to defend aggression by deliberately raising, but where possible controlling, the scope and intensity of combat', with massive nuclear strikes the ultimate threat.[97] This required the RAF to 'be capable of controlling and sustaining effective air defence operations in the face of conventional air attack as well as monitoring the integrity of the UK Air Defence Region'.[98]

However, Linesman was unsuited to a sustained conventional war. The 'centralization of information and control in the Linesman system at West Drayton in an unhardened and easily identifiable building, together with the unprotected radar sites and the exposed and inflexible broad band communications links, make the entire system extremely vulnerable to convention air attack or sabotage'.[99] The L1 Building itself was 'completely unhardened, very large and easily recognisable from the air'.[100] Nor was its radar cover suitable for a sustained conventional conflict, particularly given that low-flying aircraft designed to take advantage of ground-based radar's inability to see below the horizon were now in common usage. Linesman, an air defence study group reported in May 1972, provided radar cover that 'will not extend over the whole of the UKADR [UK air defence region] even at high level' and 'at very low level radar cover will be minimal'.[101]

Senior Air Force figures, including the Chief and Vice Chief of the Air Staff pondered Linesman's future during 1972. Although cancellation was politically unacceptable, it was not 'clear exactly what is intended for the L1 since [it was agreed] that the building could not sensibly be used as an air defence centre and

---

96   Paul Edwards, *The Closed World: Computers and the Politics of Discourse in Cold War America* (Cambridge, MA: MIT Press, 1996).

97   Final Decision on MC 14/3 A Report by the Military Committee to the Defence Planning Committee on the Overall Strategic Concept for the Defense of the North Atlantic Treaty Organization Area, 10–11. Downloaded 22 September 2014 at http://www.nato.int/docu/stratdoc/eng/a680116a.pdf.

98   'The Strike Command Air Defence Ground Environment Study Team's Proposals for the Development of the UKADGE System, May 1973: DEFE 58/114.

99   'United Kingdom Air Defence Ground Environment Systems Study Group Interim Report–1 May 1972', attached to memo from PS to VCAS, 28 April 1972. AIR 20/12429.

100   'The Strike Command Air Defence Ground Environment Study Team's Proposals for the Development of the UKADGE System, May 1973: DEFE 58/114.

101   'United Kingdom Air Defence Ground Environment Systems Study Group Interim Report–1 May 1972', attached to memo from PS to VCAS, 28 April 1972. AIR 20/12429.

there is "no merit in commissioning the RDPS'".[102] On 18 December 1972 the Air Force Board decided 'that Phase 1 of the LINESMAN project should be completed but that, after commissioning, the L1 building should be operated and manned on only a limited basis. ... Emphasis was to be placed on keeping expenditure to a minimum and high priority was to be given to the development of an alternative – less vulnerable – air defence system'.[103] The Linesman L1 building was finished by late 1973, but the system was a shadow of what was originally planned.[104]

Conclusion

Radar is an instrument that appears to produce dramatic 'time-space compression', enabling objects at a distance to be viewed as blips on a screen. In radar's early days, in the largely empty skies of World War II, such blips observed approaching the coasts of the UK could be assumed to be German (though there was confusion caused by migratory birds). By the Cold War, in an increasingly crowded sky, what mattered was not just detecting objects to provide early warning, but identifying foe from friend. Data on all known friendly aircraft was thus critical, as was the ability to process this data and make quick comparisons with the radar picture. The challenge of achieving this data handling and integration proved Linesman's downfall, not only because of unanticipated computing and software engineering problems, but also because it hinged on cooperation between divergent organisational interests and cultures. Radar was an exploratory instrument designed to scan the sky for objects, but its implementation depended on more earthbound constraints.

The instrumental correspondence between radar blips and friendly or threatening aircraft was mediated by political conceptions of what warfare might comprise in the nuclear age. Linesman took shape at a time when the nuclear threat appeared to have made air warfare obsolete. But by the end of the 1960s, Western nuclear strategists had decided that overall nuclear parity between NATO and the Soviet Union limited the utility of nuclear deterrence. So long as the US and its allies had massive nuclear superiority, it could be assumed that nuclear weapons could deter a conventional Soviet invasion of Western Europe. Once both sides had huge nuclear arsenals the threat of 'massive retaliation' came to be considered implausible as a means to deter conventional attack because all-out nuclear war would be so destructive for both sides. The solution, at least for nuclear strategists, was to argue for 'flexible response' and this required an air defence radar system that could sustain air operations during an extended conventional conflict.

102   J. A. Stocker, Secretary LINEMAN Management Board to SO to ACAS (Ops), 10 October 1972. AIR 2/18783.

103   Linesman Management Board–Minutes of the Ninth Meeting, 10 January 1973. AIR 2/18783.

104   Gough, *Watching the Skies*, 313.

The view of strategists on the relationship between nuclear weapons and the conventional military balance in middle Europe thus shaped what it meant for the UK air defence radar to work. In this sense, the politics of creating a plausible role for nuclear weapons mattered more in shaping conceptualization of time-space than any instrumental properties of radar technology.[105]

105 I am grateful to the Economic and Social Research Council for the research funding on which this chapter is based.

Index

Printed and bound by CPI Group (UK) Ltd, Croydon, CR0 4YY

22/10/2024

01777626-0019